TRACKING THE GOLDEN ISLES

Tracking the Golden Isles

THE NATURAL *and* HUMAN HISTORIES
of the GEORGIA COAST

Anthony J. Martin

The University of Georgia Press *Athens*

A Wormsloe
FOUNDATION
nature book

Publication of the book was made possible in part
through a grant from the Emory College of Arts and
Sciences and the Laney Graduate School.

Published by the University of Georgia Press
Athens, Georgia 30602
www.ugapress.org
© 2020 by Anthony J. Martin
Designed by Erin Kirk New
Set in 10 on 14 Chapparal
Printed and bound by Sheridan Books
The paper in this book meets the guidelines for
permanence and durability of the Committee on
Production Guidelines for Book Longevity of the
Council on Library Resources.

Most University of Georgia Press titles are
available from popular e-book vendors.

Printed in the United States of America
24   23   22   21   20   C   5   4   3   2   1

Library of Congress Cataloging-in-Publication Data

Names: Martin, Anthony J., 1960– author.
Title: Tracking the Golden Isles : the natural and human histories
    of the Georgia coast / Anthony J. Martin.
Description: Athens : The University of Georgia Press, [2020] |
    Includes bibliographical references and index.
Identifiers: LCCN 2019052220 | ISBN 9780820356969 (hardback) |
    ISBN 9780820356976 (ebook)
Subjects: LCSH: Natural history—Georgia—Atlantic Coast. | Barrier
    island ecology—Georgia. | Barrier islands—Georgia—History. |
    Nature—Effect of human beings on—Georgia—Atlantic Coast. |
    Atlantic Coast (Ga.)—Environmental conditions.
Classification: LCC QH105.G4 M37 2020 | DDC 508.758—dc23
LC record available at https://lccn.loc.gov/2019052220

To Richard Bromley and George Pemberton:
May your cognitive traces continue to inspire
those who read the earth and the stories
inscribed by its life.

CONTENTS

PREFACE

We live in a world defined by traces. These signs of plant and animal behavior—tracks, trails, burrows, nests, feces, rootings, borings, and much more—tell many stories about our earth both past and present, blending together to bestow a sense of the history of life while also acting as harbingers of its future. But traces do not just pass on tales. Over the past 500 million years, they also actively shaped the skin of the planet and composed entire environments, from the deep sea to mountaintops.

Where on earth would you go, then, to get a crash course in understanding the awe-inspiring and time-traveling power of traces? I suggest you take a trip to the Georgia coast and especially to take a look at its barrier islands. Although these islands are world famous mostly for reasons revolving around human affairs of a mere five hundred years or so, they deserve even more fame for their traces. Gaze on a salt marsh and behold a field of grass made possible by periwinkle grazing, fiddler crab burrowing and scraping, and mussel filtering. Stroll through a maritime forest and know the ground below teems with animal life hidden in burrows, as do trees both living and dead with their wood-boring fauna. Pass by the dunes alongside a beach and ponder how many future generations of sea turtles emerged from their sands, with mothers surviving more than thirty years at sea before coming back to those beaches and digging temporary homes for more of their kind. Scan a beach at low tide and note the thousands of tiny "volcanoes" that represent the tops of the vast tunnel systems dug by burrowing shrimp. Pick up an empty snail shell from that beach and examine its holes, healed breaks, and other markings to learn not only that it lived its life and died but also that its shell hosted other lives that went on epic journeys.

Curious? Regardless of whether you can or cannot physically go to the Georgia sea islands, this book—*Tracking the Golden Isles*—is intended to give a few appetizers of the traces there and the stories they tell. Having spent a good chunk of my life studying and teaching about plant and animal traces of the Georgia coast, I tried to share some of what I learned with others through a previous book, *Life Traces of the Georgia Coast* (2013). Yet after four years of working on that book, and as pleased as I was once I felt its considerable weight in my hands, it still felt oddly incomplete. More accounts were out there being written in the sand, mud, wood, shells, and bones of the islands, and each time I went to look for them, there they were, and they never failed to confer new insights. So in 2011, even before the predecessor to this book was published, I began writing some of these new lessons in a blog bearing its name, *Life Traces of the Georgia Coast*. Much of the book you hold now was birthed through that blog, an eight-year gestation that also includes updated information and new musings about the gorgeous natural places and biota of the Georgia coast.

Nonetheless, one of the realities of the coast I also try to convey with this book is how it is not all just cordgrass and crabs, and it is not a perfect time capsule of what once was. Because people have been a part of the islands and their ecosystems for more than four thousand years, these places are a broken mirror in which we can admire their beauty while also acknowledging their flaws. The first people on the islands, the probable ancestors of the Guale and Timucua, came voluntarily, and they left traces of their culture and cooperation. Much later, representatives of powerful institutions from Europe arrived, regimes that then waged wars of conquest and cultural assimilation that erased most of those who preceded them. Other people were later captured, enslaved, and forcibly shipped across an ocean, where the survivors of those brutal journeys worked island lands for the commercial benefit of the colonizers. Even after earning their freedom, these people, their former enslavers, and their descendants continued to change the islands and other nearby environments up to now and will continue to do so into an uncertain future. The newcomers also brought plants and animals that had evolved elsewhere but soon displaced native species and altered entire habitats.

All the while, though, ghost crabs punched holes into coastal dunes; woodpeckers drilled into live oaks and cabbage palms; raccoons dug up, ate, and defecated the remains of fiddler crabs; moles burrowed for and found earthworms to their liking; moon snails hunted hapless clams; and ghost shrimp plumbed deeply into offshore sands. No matter what happens next in our relationship with the Georgia sea islands and the uncertainty of massive changes already underway, we can be assured that traces remain and traces will be made, with new stories in process.

No book, whether about islands or not, is an island unto itself. So before celebrating scripts, human or otherwise, I must thank all who made this book possible. First, I appreciate my agent, Laura Wood (FinePrint Literary), for supporting my writing this book and for affirming University of Georgia Press as my first choice for a publisher. As a graduate of the University of Georgia (PhD 1992), writing for UGA Press felt like returning to my intellectual home and a rightful way to honor the place where I first learned about the Georgia coast.

Second, I want to thank the University of Georgia Press staff, particularly acquisitions editor Patrick Allen. In a geographically appropriate way, I pitched my idea for this book to Patrick in January 2018, as we both stood on Jekyll Island (Georgia) at the One Hundred Miles' Choosing to Lead conference there. My gratitude also goes to other UGA Press staff: chief editor Lisa Bayer, for approving the book proposal; assistant to the director and editorial assistant Katherine La Mantia; intellectual property manager Jordan Stepp, for composing the book contract; and Jon Davies (project editor), Erin New (designer and compositor), and Melissa Buchanan (production coordinator), for helping make this book real. Two external reviewers—one anonymous, and one not (thank you, Dr. Evelyn Sherr)—provided helpful suggestions for improving my original proposal for the book and later convinced me of what should stay and what should go. The copyeditor, Susan Silver, was invaluable for polishing my prose and otherwise detecting oddities, inconsistencies, and just plain dumb mistakes, ensuring a more readable book. Of course, if you see anything amiss, factually or otherwise, that's on me.

In my forays to the coast, I have been extremely fortunate to know some of the best scientists, naturalists, historians, and other knowledgeable people associated with the Georgia coast. Because I regrettably do not live on or near the coast, nor do I stay there long enough to gain a truer sense of place, I rely on their expertise to fill the gaps of what I miss during the times between. With the understanding I will somehow forget someone (apologies in advance to you), the ones I recall for now include, in alphabetical order, Merryl Alber (UGA Marine Institute); Clark Alexander (Skidaway Institute of Oceanography); Craig Barrow (Wormsloe Historic Site); Gale Bishop (Georgia Southern University); Jim Bitler (Ossabaw Island); Tim Chowns (University of West Georgia); Scott Coleman (Little St. Simons Lodge); Melissa Cooper

(Rutgers University); John "Crawfish" Crawford (University of Georgia Marine Extension, Skidaway Island); Edda Fields-Black (Carnegie-Mellon University); Christa Frangiamore Hayes (Coastal Wildscapes); Robert W. "Bob" Frey (University of Georgia); Mark Frisell (Ossabaw Island Foundation); Jon Garbisch (formerly UGA Marine Institute); Yvonne Grovner (Sapelo Island); Robin Gunn (Ossabaw Island Foundation); Ann and Andrew Hartzell (Savannah); Fred Hay (Sapelo Island); Royce Hayes (St. Catherines Island); Steve and Kitty Henderson (Emory University); Stacia Hendricks (Little St. Simons Island Lodge); V. J. "Jim" Henry (Georgia State University); Jen Hilburn (Altamaha Riverkeeper); Brian Meyer (Georgia State University); Tiya Miles (University of Michigan); Steve Newell (Jekyll Island); Janisse Ray (Reidsville, Georgia); Jim and Shelley Renner (St. Simons Island); Sarah Ross (University of Georgia, Wormsloe Foundation); Carol Ruckdeschel (Cumberland Island); Katy Smith (Georgia Southern University); Mart Stewart (Western Washington University); David Hurst Thomas (St. Catherines Island); Gracie Townsend (UGA Marine Institute); and R. Kelly Vance (Georgia Southern University). Three of these people—Jim Bitler, Bob Frey, and Jim Henry—have departed this existence since they taught me a little of what they knew. I'll do my best to pass on those pieces before I go too.

Part of my incentive for writing this book also stemmed from a grand project about the coast at my home institution (Emory University) and in cooperation with the Center for Digital Scholarship, called the *Georgia Coast Atlas*. Intended as a free online source of information about the Georgia coast for the general public and scholars, we started the atlas in September 2015 and plan to continue it well into the future. Colleagues, teammates, and friends involved with the project on- and off-campus are Steve Bransford, Anandi Salinas Knuppel, Wayne Morse, Shannon O'Daniel, Michael Page, Rebecca Page, and Allen Tullos. Thank you for your support and for helping me to become a better public educator.

In my study of ichnology, the study of traces, I am privileged to have learned from some of the best ichnologists and paleontologists in the world. These include, in alphabetical order, Richard Bromley (University of Copenhagen); Luis Buatois (University of Saskatchewan); Kathleen Campbell (University of Auckland); Al Curran (Smith College); Tony Ekdale (University of Utah); Jorge Genise (Museo Argentino de Ciencias Naturales Bernardino Rivadavia, Buenos Aires); Jordi Gilbert (University of Barcelona); Murray Gingras (University of Alberta); Roland Goldring (University of Reading);

Murray Gregory (University of Auckland); Patricia Kelley (University of North Carolina-Wilmington); Martin Lockley (University of Colorado–Denver); Gabriela Mángano (University of Saskatchewan); Radek Mikuláš (Geologicky´ ústav AVČR, Czech Republic); Renata Guimarães Netto (UNISINOS, Brazil); George Pemberton (University of Alberta); Thomas Rich (Museum Victoria); Andrew Rindsberg (University of West Alabama); Dolf Seilacher (University of Tübingen); Alfred Uchman (Jagiellonian University); Patricia Vickers-Rich (Monash University); Sally Walker (University of Georgia); and Andreas Wetzel (University of Basel), to name a few. This list also includes a few who left us but whose teachings stayed with me: Roland, Jordi, Dolf, George, Richard, and Murray. You gave me much to ponder, and I am better for it.

A shout-out is warranted to the organizations that work tirelessly to protect and educate about the environments on and along the Georgia coast. (Actually, I'll bet they get a little tired.) These organizations include Altamaha Riverkeeper, Caretta Research Project, Center for a Sustainable Coast, Coastal Conservation Association, Coastal Wildscapes, Cumberland Island Conservancy, Driftwood Education Center, Friends of Sapelo, Georgia Conservancy, Georgia Sea Turtle Center, Glynn Environmental Coalition, Initiative to Protect Jekyll Island, Jekyll Island Foundation, Manomet, One Hundred Miles, Ogeechee Riverkeeper, Ossabaw Island Foundation, Satilla Riverkeeper, Savannah Riverkeeper, Sierra Club (Coastal Group), St. Catherines Island Foundation, St. Simons Land Trust, Southern Environmental Law Center, Stewards of the Georgia Coast, Nature Conservancy (Coastal Office), Tidelands Nature Center, Tybee Island Marine Science Center, University of Georgia Marine Extension Service, and the Wormsloe Institute for Environmental History. Keep up the good fight.

My heartfelt appreciation also goes out to what are now generations of undergraduate students at my home institution of Emory University. Through field trips to the coast and many classes I have taught for thirty years, my students remind me that education is a two-way street, in which the world is a classroom where a heightened awareness delivers a constant flow of new lesson plans. This book honors them and their future contributions to science and society.

Next to last but not next to least is my wife, Ruth Schowalter. She is my companion, muse, colleague, great encourager, artist, and best friend through all the times we have spent on the Georgia coast together. She also continues to support my annoying habit of writing books in between living the rest of our wild, precious lives. Thank you for all past, present, and future times there, Ruth.

No acknowledgments for this book would be complete without also pointing toward those that made it truly possible, the myriad of tracemakers on the Georgia coast: the plants, insects, spiders, earthworms, clams, snails, crayfish, shrimp, horseshoe crabs, hermit crabs, actual crabs, amphibians, reptiles, birds, mammals, and more. Their traces preceded us there, and some no doubt will exceed us. Let us cherish our great luck in sharing the same sliver of geologic time with these beings and wonder at their handiwork.

Part I

Impressions of Past Histories

1

Knobbed Whelks, Dwarf Clams, and Shorebirds

The beach was teeming with predators. We did not know this before setting out on our bicycles early that morning, and we probably passed over many of these carnivores before recognizing their distinctive and undeniable outlines under the sand. Once these patterns announced the hunters' presence, we understood that they had willingly buried themselves, a waiting game of survival in acquiescence to waning waters. What made this situation even more remarkable, though, was how their submergence also unwittingly supplied temporary refuge for another, much smaller, species—that is, before other predators arrived on the scene and feasted on each of the smaller species as hapless prey.

My wife, Ruth, and I were lucky enough to witness the results of this complicated life-and-death beach drama because we were on vacation. For nearly every Thanksgiving since 2010, instead of giving thanks for family, football, and native fowl, we pay homage to the Georgia coast by placing our feet on its sandy shores. To make this nature therapy happen, we flee the metropolitan Atlanta area either the Wednesday before or the morning of the last Thursday of November and then drive about five hours before stopping on the barrier island of Jekyll.

Jekyll Island is an odd place. It is one of more than a dozen barrier islands on the coast of Georgia, its southern tip less than a marathon distance north of the Georgia-Florida border. Unlike most barrier islands in the eastern United States, the Georgia barrier islands are relatively undeveloped, with the majority hosting lush maritime forests, long and wide sandy beaches, coastal dunes adorned with swaying sea oats (*Uniola paniculata*), and vast expanses of salt marshes, with few humans. Jekyll, however, is one of the few Georgia barrier islands connected to the mainland by a causeway, allowing car-bound visitors to easily arrive, depart, or pause. For those who choose to stay, it hosts paved roads, neighborhoods, golf courses, boutique shops, and beachside hotels, but not so densely packed as to invoke nightmarish visions of *Jersey Shore*. Jekyll also has paved bicycle paths around the island that wind through maritime forests and salt marshes or occasionally slip behind coastal dunes for stunning views of the Atlantic Ocean. Even better, if you have your own bike and are hankering to go on paths less traveled, Jekyll offers long stretches of sandy beaches, especially on its south end. These beaches widen considerably at low tide, their smooth, flat, hard-packed quartz-sand surfaces beckoning riders to give them a try.

Fittingly, then, we pointed our bikes south along the beach our first morning there, leaving hotels and condominiums behind. During this exhilarating outing—breathing in and feeling the cool, salt-infused air passing over exposed skin; delighting in the gentle lapping of waves; and listening to the far-off calls of gulls—Ruth and I stopped occasionally. These breaks were not so much for rest but for science, giving us a better chance to look at and learn from any animal traces—tracks, burrows, trails, and more—that captured our curiosity. We were practicing the age-old science of ichnology, the study of traces. It is observation done with a low-carbon footprint, natural history that is also eco-chic. Because we had been to Jekyll enough times to know where its best traces are likely to be found, we used our insider knowledge to focus on all things ichnological. This forethought also means we sometimes discovered phenomena that, as far as we know, were previously unnoticed on any of the Georgia barrier islands, including the more natural ones lacking posh hotels and miniature golf.

It was during one of these stops that we discovered the signs of hidden predators and other animals leaving mysteries for us to intuit. The main cast of characters in that then-novel discovery included two molluscans, knobbed whelks (*Busycon carica*) and dwarf surf clams (*Mulinia lateralis*); and two species of shorebirds, sanderlings (*Calidris alba*) and laughing gulls (*Leucophaeus altricilla*). How these four animals and their traces related to one another made for a fascinating story, nearly all of it discerned through their traces left on that Jekyll Island beach. Later I realized a fifth species— tiny crustaceans sometimes nicknamed "sand fleas" but what scientists prefer to call amphipods—must have also played an indirect role in the marvels we observed during that visit and never since.

We first spotted the traces at low tide and farther downslope on the beach, next to the rhythmic and comforting swashing of waves. There we noticed upraised and cohesive flaps of sand that looked like triangular "trap doors." If you ever find one of these, gently place your fingers down and underneath a few centimeters, lift up, and you will be holding a living whelk—probably a knobbed whelk. Knobbed whelks, which are large, thick-shelled marine snails, are the most common of three species of whelks on the Georgia coast.[1] Because of the knobbed whelk's size and beauty—bearing prominent studs on its widest whorl, grayish-white longitudinal stripes, and an orange interior—it is perfectly understandable why it is also the official state shell of Georgia.[2]

Human-bestowed admiration aside, knobbed whelks are predators that seek out, kill, and eat clams, oysters, and other bivalves.[3] To do this they use the edges of their robust shells to wedge apart appropriately sized living shells. Once they succeed in exhausting these recalcitrant bivalves, they insert their proboscis, a feeding apparatus that includes its mouth, esophagus, and a hard, rasping tongue called a radula. The radula, which is a structure made of the same material as its shell, is used to scrape and consume a clam's soft innards. More often than not they do their killing offshore, well away from the prying eyes of beachcombers. Yet sometimes their handiwork washes up on shore for us to find as empty hinged clamshells with chipped margins.

These particular knobbed whelks had been brought in by strong waves with a high tide about six hours before we rode on the beach. Once the waves subsided and the tide dropped, the whelks burrowed down into the sand. There they would wait until the next high tide, and with its advance they would move out of their temporary shelters and go back to crawling and predating. This tactical behavior of whelks, which avoids both desiccation and predation by others, has been positively reinforced by millions of years of natural selection in their lineage; hence most are quite good at it.[4]

Considering that whelks lack both shovels and arms, how do they bury themselves? They use a muscular foot, expanding and contracting it to displace still-saturated soft and pliable sand left by a high tide. Once a whelk foot protrudes from its shell and gets far enough into the underlying sand, it anchors there and pulls the rest of itself sideways and down.[5] This is not so much burrowing as it is intrusion, where the animal insinuates itself into the sand. Contrast this method with the active digging we normally associate with most land-dwelling burrowers, many of which are aided by legs and leave open holes in the ground as obvious products of their tunneling.

Once the whelk is buried, waves wash over its trail, erasing all evidence of its most recent activity. Nonetheless, once the sandy surface is emergent for an hour or so, seawater draining downward through the sand tightens grains around the whelks, defining them as the previously mentioned trap doors. Such outlines may also bear a small hole at one end of the triangle, marking where the whelk expelled water from the bottom of its shell. All these clues state clearly that this animal is still very much alive and not a shell to take home and put on a shelf—that is, unless you enjoy large marine gastropods crawling around briefly before dying in your home.

Mystery solved, right? Not quite. Near these clear examples of whelk traces on the beach were clusters of dwarf surf clams. Similar to whelks, these yellow to off-white clams were washed up by the high tide and instinctually burrowed once exposed on the surface. Surf clams are much smaller and more streamlined than knobbed whelks, about the size and shape of almonds, roasted or otherwise.[6] But they likewise use a muscular foot to intrude the sand, anchor, and pull in their shelled bodies. Under the right conditions these clams will also leave surface trails behind them before descending under the sand, although such paths are easily wiped clean by a single wave.[7]

Dwarf surf clams ideally orient themselves vertically and push two siphons through the sand, one of which sucks in seawater bearing suspended food particles and the other to expel wastes. The siphons make paired holes visible on sandy surfaces, and the clam body below these completes a Y-shaped burrow.[8] Sometimes, though, dwarf surf clams have only enough time and gumption to bury themselves on their sides, concealed by a mere cap of sand. This clam equivalent of hiding under a blanket makes them much more vulnerable to predation, especially from shorebirds that find them and commence snacking.

Among such voracious shorebirds are sanderlings.[9] Sanderlings are exceedingly common on Georgia-coast beaches all year, which is a good thing, because they are adorable little white-brown shorebirds that also run at amusingly cartoonish speeds along shorelines. Whenever not running or flying, sanderlings are sticking their beaks into the sand and snatching up small animals below the surface, including dwarf surf clams. Sure enough, if you find a cluster of these clams, you will also likely find abundant small three-toed sanderling tracks and beak probes, the latter looking like someone poked a pencil rapidly and repeatedly into the sand.[10] Their food choices

FIGURE 1. Whelks playing hide-and-seek and getting attention for it. *Left*, a knobbed whelk exposed at low tide on a Jekyll Island beach tries to bury itself, biding time until the next high tide. *Right*, a successfully buried whelk attracts a following from a bevy of bivalves, the dwarf surf clam.

are clarified even more when you see their tracks and beak probes directly associated with eyelike holes where they neatly extracted little clams from their burrows.

So how did these three species and their traces all relate to one another? This is where matters got even more interesting. In the center of the clam clusters were bare spots on the sand devoid of both clams and beak pokes, and they had triangular outlines. Underneath these outlines were whelks. As Ruth and I stood back and looked down the beach, we marveled at how these distinctive clumps of clams were spread throughout the exposed sand-flat, with each clump surrounding a whelk. Somehow the whelks served as attractive sites for the clams, which chose to burrow in the sand around them, rather than randomly dispersing themselves throughout the beach.

Why were the clams choosing to burrow around the whelks? Was this some sort of commensalism, in which the clams found more food around the whelks? No, because these clams are filter feeders, using siphons to take in water with bits of organic matter for their sustenance, and at low tide they would be filtering only air, which is not so nutritious. How about protection? That did not seem likely either, because the whelks had no interest or reason to defend the clams, and their bodies were not serving as living shields against shorebirds.

So I thought about how these clams burrow, and then it all made sense. Because dwarf surf clams are so small, sand grains to them are more like pebbles to you and me. Moving their bodies through these sediments thus takes considerable effort, especially as water drains from the sand and surface tension binds grains together more tightly. This means the clams take advantage of sand that acts more like quicksand and less like concrete, and the burrowing is best when the sand still has plenty of water between grains.

This is where each whelk became both the unwitting friend and enemy of the dwarf surf clams. As a whelk burrowed, it fluidized the surrounding sand, shaking up the grains so that more space opened between them, which allowed in more water. This zone of disturbance and liquefied sand was eagerly exploited by nearby clams, which burrowed more easily into both the whelks' trails and the immediate areas around their bodies. Ruth and I later verified this interpretation when we found a few whelks caught in the act of burrowing, with bunches of surf clams in their wake.

Alas, these safe spaces temporarily provided by the whelks ultimately led to the sanderlings chowing down on the clams, transforming what might have been a meticulous search for morsels sprinkled hither and tither throughout a Jekyll Island beach to instead become an effortless meal. Now all a wandering sanderling had to do was find a motherlode of molluscan goodness conveniently concentrated around a buried whelk and start probing. It was an all-you-can-eat clam feast, and the traces clearly showed where some of these birds stopped and took their time gorging on clams. In a few instances their tracks also showed where one binging sanderling would attract the attention of others, who rushed to the scene and joined in the buffet.

Was this the only trace-enhanced form of predation taking place on that beach? By no means, and it was not even the only one involving whelks and their traces, as well as sanderlings getting a good meal from someone else's traces. This is where a new character—the laughing gull—and a cast of thousands of amphipods enter the picture. Considering that knobbed whelks are so large and meaty, it only makes sense that a bigger animal would want to eat one whenever it washes up onto a beach. Indeed, gulls add

FIGURE 2. Signs of carnage in the wake of a roaming pack of sanderlings, aided by a burrowing knobbed whelk. By following the whelk as it burrowed and by clustering in such a small area, the dwarf surf clams unknowingly hastened their doom.

knobbed whelks to their already-lengthy menus. Hence when a flying gull sees a whelk doing a poor job of playing hide-and-seek during low tide on a beach, it will land, walk up to the whelk, and pull it out of its resting spot. From there the gull will either consume the soft insides of the whelk on the spot, fly away with it to eat elsewhere as a take-out meal, or reject it, leaving it high and dry next to its original resting trace.

All these scenarios result in variations of a vestige theme. First, you see an offset pair of webbed-foot tracks with skid marks, showing where a gull in flight landed on the beach. Then you see its narrowly diagonal alternating gait—left, right, and left—stopping with side-by-side tracks in front of a triangular impression. This place shows exactly where the gull positioned itself to bend over and pick up the narrow end of a whelk with its beak. The triangle in the sand next to these tracks may be fairly complete, telling of a whelk that only had time to partly bury itself. But the outlines of these depressions also might be messy, with a fractured flap of sand showing how gulls can rival humans by recognizing buried whelks and ripping them from their hiding places. An additional gull-predation trace might be formed elsewhere, when gulls take off with the whelk in their beaks, carry them up, and drop them onto hard-packed beach sand to crack open their shells, revealing their yummy interiors.[11]

Like all good scientists, Ruth and I tested this gull-whelk predation hypothesis. On the same Jekyll Island beach that morning, Ruth and I looked for and found impressions where whelks covered themselves at low tide, only to be pried out by laughing gulls. Although gull tracks tend to look alike, with three forward-pointing webbed toes in front and a little nub of one in the rear, different species can be distinguished by size.[12] For instance, herring gulls (*Larus argentatus*) are monstrously huge, so their tracks are accordingly the largest gull tracks anyone will find on a Georgia beach. Laughing gulls and ring-billed gulls (*Larus delawarensis*) are similarly sized, but laughing gulls are slightly smaller and make appropriately shorter and narrower tracks. The footprints we found that morning around the whelk traces were those of laughing gulls, which we also saw and heard later. Nearby were landing tracks, and we even occasionally found a discarded (but still live) whelk bearing the same dimensions as the impression next to it as a facile way to test our presumption. In short, we did not need to witness laughing gulls preying on and prying up whelks; we knew what happened because of the traces.

FIGURE 3. *Left*, a knobbed whelk feeling exposed and rejected after having been pulled out of its hiding place by at least two laughing gulls, their identities revealed by their tracks. *Right*, resting trace of a whelk, but one nearly erased by sanderlings trampling and pecking, telling of an amphipod buffet there.

But some vaguely defined triangular impressions with blurry outlines presented another little mystery. A closer look revealed these were also resting traces made by whelks, but ones nearly obliterated by sanderling tracks and beak marks. With these there were no recognizable sign of gulls having been there, nor were any whelk bodies in sight, nor any surf clams. Perhaps, then, these were instances of where the gulls flew away with successfully acquired whelks to drop and eat somewhere else. Still, why did the sanderlings follow up on the gulls with these shorebirds throwing a big party in a small place, especially if nobody brought the clams?

Although many people may not be aware of this, when they walk hand in hand along a sandy beach on Jekyll or anywhere else on the Georgia coast, a small-shorebird smorgasbord lies under their feet in the form of amphipods. Through sheer numbers amphipods can compose more than 20 percent of the animals living in Georgia beach sands, capable of providing about 10 percent of shorebird calories.[13] These tiny crustaceans normally spend their time burrowing through beach sands and eating microscopic algae between sand grains or on their surfaces. This collective action also constantly shifts billions of individual sand grains, a cryptic mass movement of sediments that to our eyes seem completely at rest.[14] Because amphipods are exceedingly abundant and just below the beach surface, they represent a rich source of protein for small shorebirds. But if you really want to help shorebirds get at this food, just drag your feet as you walk down the beach. This action will expose the crustaceans, and shorebirds will snap up many of them before they can burrow back into the sand.

This concept of amphipod exposure leading to sanderling feasting came to Ruth and me during another visit to Jekyll, but on its north end. Tourists on an early morning horse ride along a beach there unwittingly fed the sanderlings, which swooped in and selectively mined the horse tracks for food. And this, I think, is also what happened with the gulls pulling whelks off the beach surface. The seemingly simple one-on-one act of a gull picking up a whelk also exposed enough amphipods to attract sanderlings, which then set off a predator-prey interaction between the sanderlings and amphipods, centered on the resting trace of the whelk.

Given such insights, this is about when I take off my newish, nylon-fabric, broad-brimmed beach hat and instead put on my dusty, old, leather paleontologist hat and ask myself a few questions. One is this: if I found a concentration of small clamshells around a big snail shell in an ancient beach deposit, might this clustering have been the result of an enabling snail loosening the beach sand for the clams? Yes, and this would be easy to discern. Second, would I then connect three-toed tracks, holes made by beaks, and holes outlining the bodies of missing clams as evidence for voracious shorebirds feeding on the clams? Sure, I would get that too, and so would most paleontologists once they had the right search images and visions.

But the last set of traces would be more challenging. With this I might see a sandstone surface bearing a dense accumulation of overlapping three-toed tracks—and with only a few clearly defined—on an otherwise irregular surface riddled by shallow holes. The triangular depression marking the former position by a large snail, obscured by hundreds of tracks and beak marks, might stay unnoticed or, if seen, could be disregarded as an errant scour or some other mark devoid of life-generated intent. The larger three-toed tracks made by a gull or similar larger predator would be gone, overprinted by the many footprints and beak impressions of the smaller birds. Hence the role of the instigator for this chain of events—the gull or its paleontological doppelgänger, as well as its large prey item—would remain both unknown and unknowable. It is a humbling thought, and it exemplifies how much geologists, paleontologists, and others who study prehuman pasts should wonder what they miss when they attempt to recreate ancient worlds without the imagination-boosting lessons of today.

Another message delivered by these ichnological musings is that all the preceding observations and ideas blossomed from one morning's spouse-aided bicycle ride on a Georgia coast beach. Even more noteworthy, these interpretations of natural history were made on an island that some of my fellow scientists might dismiss as too developed to study, its biota and their ecological relationships somehow sullied or tainted by a human presence. How do I know this? I was one of those snobs. But no more: now I look for natural histories wherever they might be, no matter how much they have intertwined with human histories.

How these histories all come together through the life habits and traces we and other animals leave behind illustrates how the Georgia coast offers universal teachings in how the consequences of behavior can be more important than the behavior itself. In this instance the traces of only a few animals triggered changes in the behavior of other animals, sometimes helping them and sometimes hastening their demise. These manuscripts in the sands, muds, shells, and bones of the Georgia barrier islands help to clarify complicated relationships, showing how seemingly inert tracks, trails, burrows, and other traces can sway decisions, impinge on individual lives and entire ecosystems, or encourage unintentional and seemingly unlikely partnerships between species.

These revelations on and following a Thanksgiving holiday also led me to extend a genuine wish to all who visit or otherwise know of the Georgia barrier islands. Let us all be thankful for the natural areas still preserved on Jekyll Island and other parts of the Georgia coast that allow for such wanderings of our bodies and minds, and let us be thankful for little personal discoveries of their life traces, infused with wonder and shareable with others.

2

The Lost Barrier Islands of Georgia

My first visit to the Georgia coast was on Sapelo Island, and I was there because of its geology. Yet all of my geologically oriented memories from that trip were displaced, which is what happens when you witness an alligator (*Alligator mississippiensis*) kill a cocker spaniel (*Canis lupus familiaris*). This unfortunate incident was made possible by a visiting researcher there who inexplicably allowed his dog to run free and unsupervised in a place with a freshwater pond nicknamed the "alligator pond." The pond's appellation was further justified by a large mother alligator lying on a bank next to the pond, where she watched over more than a dozen of her 'gator babies. The spaniel harassed these hatchlings for the first two of three days I was there, jumping into the water and swimming after them. His last leap was on the third day, when the mother was off the bank and in the water, waiting for the harasser of her children. Bubbles followed the big splash, and a little less than an hour later, I had to deliver the bad news to the researcher. It was not pleasant.

This sort of event was an early lesson for me that predation is a fact of life on the Georgia coast and that unnatural selection happens when newly specialized dog breeds encounter predators honed by millions of years of evolutionary history.[1] Yet it also effectively distracted from anything I might have learned then about the geologic processes of Sapelo Island. Not until much later did I realize that the intertwined products of geologic and biological processes there and on other Georgia barrier islands teach us how most of these barrier islands long preceded the advent of people in North America, let alone their ill-fated pets.

My witnessing of canine demise happened in 1988 while I was a PhD student at the University of Georgia, and I was on Sapelo Island specifically because of a three-day class field trip with other graduate students. The class was titled Advanced Sedimentation but also could have been called Sedimentology, which is the study of sediments, such as mud, sand, and other unconsolidated earth materials; hence it is a small part of geology. But sedimentology also involves discerning how sediment moves and is deposited, as well as how structures and sedimentary rocks formed by these processes inform us of past environments.[2]

For example, most of the expansive and gasp-inducing beaches of the Georgia coast are composed of very fine to fine-grained sand primarily composed of the mineral quartz but also including a smattering of darker and heavier minerals, such as rutile (a titanium oxide) and ilmenite (an iron-titanium oxide).[3] To a romanticist a beach is a rapturous place, inviting activities ranging from competitive sports to cherished idleness. To a sedimentologist, though, a sandy beach is a mystery begging to be solved, provoking many questions, such as, Why is the sand mostly quartz instead of other minerals? Where did this sand come from? How did it get onto these beaches? And why is the sand on the open-ocean side of the barrier islands, instead of behind them in the marshes and sounds, which are instead smothered by thick slurries of mud? Interacting with real sediments in their places of deposition rather than reading about them in all-too-clinical journal articles more readily answers such inquiries. On field trips you can feel the sand with your fingers, sniff it, get it into your clothes, listen to it squeak underfoot, or otherwise get to know it much better than through mere words.

In terms of the history of science, the Georgia coast is known for its contributions to sedimentology but is more famous for its historical role in the development of modern ecology. Since the 1950s ecologists—people who study the connections between living and nonliving things in ecosystems— can lay claim to Sapelo Island as the birthing place for what are now mainstream concepts in ecology, such as nutrient cycling and transference of matter and energy in ecosystems.[4] But what about geologists? Fortunately, they were not long behind the ecologists, beginning their research projects on Sapelo Island and other Georgia barrier islands in the 1960s. From that original work and investigations that continued for decades afterward, these islands are now renowned for the insights they bestowed on our understanding of sedimentary geology and especially for where ecology intersects with sedimentology.[5]

You may be wondering why geologists—infamous for their obsessions with hard objects—would be attracted to these islands made of shifting sand and mud nearly bereft of anything resembling a rock. Part of this question is answered by an awareness that sediments come from rocks, meaning any discussion about the birthing of sediments also necessitates talking about their parentage.

Rocks are classified by three categories: igneous, metamorphic, and sedimentary. Igneous rocks are the products of molten rock (magma) cooling and solidifying, whether below the earth's surface (intrusive) or on its surface (extrusive), such as granite and basalt, respectively.[6] Metamorphic rocks form when preexisting rocks are put under pressure, such as during mountain building, or when they are subjected to heat, which happens if those rocks are adjacent to hot magma. But the heat cannot be so much that it causes melting, because this rock then enters the igneous realm. Sedimentary rocks are made from sediments, and sediments are produced when any rock—igneous, metamorphic, or sedimentary—is broken down at the earth's surface and later reformed into a new rock. Sedimentary rocks are in turn placed in two categories: chemical and clastic. Chemical sedimentary rocks are a result of minerals precipitating out of water, which includes all limestones, rock salt, and more. In contrast, clastic sedimentary rocks are composed of broken bits and pieces of other rocks, especially those that mostly have silica-bearing minerals, such as quartz. Examples of these rocks include sandstones, siltstones, and claystones.

Now if the professor of my graduate-school geology class had decided that her students needed to learn about the formation of sediments making up chemical sedimentary rocks, she would have taken us to the Bahamas. (Yes, geologists sometimes go to very nice places for their field research but later endure grief from people who always assume such work is a vacation.) Still, for learning about clastic sedimentary rocks, few places in the world are better than the Georgia coast. Three decades later—as a geologist who has seen clastic sedimentary rocks in nearly forty states of the United States and more than twenty countries outside the United States—I still draw on my experiences from the Georgia coast as a reliable standard for interpreting these rocks. Other geologists feel the same, and more than a few have expressed envy at my good fortune in having spent so much time on the Georgia barrier islands. Thus I can reasonably claim that the sands and muds of the Georgia coast, despite their squishiness, have made me a much better geologist.

As mentioned previously, sediments are needed before sedimentary rocks can be made, and those sediments must get deposited before solidifying into rock. So geologists interested in learning how the modern sands and muds of the barrier islands are deposited, eroded, or otherwise moved in coastal environments can watch and study these processes every day along any Georgia shoreline. The products of sediment movement and deposition are sedimentary structures, most of which are either from physical processes— such as winds, waves, or tides, forming ripples and dunes—or biological processes, such as burrowing.[7] Accordingly, sedimentary structures are clas-sified as either physical or biogenic, respectively. (A third category—chemi-cal sedimentary structures—is duly acknowledged but is not relevant here, so you have my permission to forget about it for now.)

Yet another important part of sedimentology is knowing that if you ever have a conversation with a sedimentologist at a cocktail party, wedding, or (most likely) dive bar, make sure you know the difference between clay, mud, silt, sand, pebbles, cobbles, and boulders. What I just listed are sizes of sed-iments and in order from smallest to largest. Sediment size is part of how sedimentologists describe sediment texture, which is the size, shape, and arrangement of grains.[8] Take sand. I always implore my students to think of "sand" as a size term regardless of its composition, as it is applied to par-ticles $\frac{1}{16}$ to 2 millimeters (0.0025 to 0.08 inches) wide. Sand is then further subdivided into very fine, fine, medium coarse, and very coarse. Sure, not all sand is the same. For instance, Bahamian beach sands are made of cal-cium-carbonate minerals, whereas Georgian sands are mostly quartz. Still, they are united by size. Any sediment larger than 2 millimeters is no longer sand but a pebble, cobble, or boulder, and anything less than $\frac{1}{16}$ millimeter is either silt or clay. Typical beach sand on an undeveloped Georgia barrier island, such as Sapelo, St. Catherines, Ossabaw, or Cumberland, is very fine to fine grained, which is $\frac{1}{16}$–$\frac{1}{8}$ millimeters (0.0025–0.005 inches) wide.[9] If you do not happen to have Georgia beach sand handy just now, you can vi-sualize this size range by imagining all the periods in this book in the palm of your hand, and for the larger sand sizes do the same with periods from large-print children's books.

Georgia salt marshes, in contrast, are dominated by clay and silt-sized particles, which when mixed together with water are collectively and simply called "mud." For most people "clay" is a fairly straightforward word. But it is potentially confusing in sedimentology because it refers to both a group of minerals with a sheet-like structure—appropriately called "clay miner-als"—and sediments less than $\frac{1}{256}$ millimeter in diameter.[10] No matter how hard you squint, you will not be able to see clay-sized sediments without using a powerful magnifying glass or a microscope. Silt can be seen, but not easily, and is better detected by touch. One geology professor of mine taught me that the quickest way to tell the difference between a claystone and a siltstone is to rub it against your teeth: if it feels smooth, it's claystone, but if it feels gritty, it's siltstone. (He also recommended, however, that geologists not conduct this test so often that they wear down their teeth over a lifetime of rock identification.)

Size is arguably the most important aspect of a sediment texture, but what about shape and arrangement? Shape is important for estimating how far a clastic sedimentary particle has traveled from its original source rock. For instance, a well-rounded and spherical grain of quartz sand has very likely moved much farther than, say, an angular and oblong boulder of quartz. (Incidentally, boulders are defined as sediments that are more than 256 millimeters [10.1 inches] wide, but please do not try to explain this to someone while warning them of a falling boulder.) This general rule implies that clastic beach sand may have gone a long way before making its way to a beach and into the cracks and crevices of your body. Arrangement refers to how sediments are packed: loosely with grains barely touching one another, tightly with lots of contact points, or something in between those two. This means that when you step on a beach, each footprint alters its texture.

How all these literally nitty-gritty details relate to rocks was intimated earlier by my mentioning the rock names of sandstone, siltstone, and claystone, as well as sedimentary structures. How does sediment become a rock? Sediments are bound together by hardening of either clay minerals (matrix) or cement. The latter is similar to the cement in a typical sidewalk but refers generally to any minerals that precipitate from a watery solution. Think of how a rippled sandy beach at low tide, or the wavy sand dunes just up the beach from these, might be "frozen" in place with just a little bit of matrix or cement between the grains. Such a magical transformation would then preserve sediment textures of those originally sandy environments. (But textures in mud change majorly over time, as explained later.) It would also preserve the physical sedimentary structures, such as layering—which geologists call "bedding"—as well as ripples, dunes, burrows, tracks, root traces, or other biological disturbances of the sediment.

In most instances, though, sedimentary rocks need lots of time to become real rocks and are not quite there yet. For example, a series of northeast-southwest trending ridges on the lower Georgia coastal plain—most within biking distance of the present-day Georgia coast—are still just sandy rather than sandstone. As a result, any highway commission tasked with cutting a new road through these ridges would use backhoes and bulldozers rather than drill holes and dynamite. Despite how the sandy ridges farthest from the Georgia coast are perhaps a couple of million years old, they have not yet experienced the right geologic conditions for their loose sand grains to cement. Nonetheless, these ridges preserve the sediments and sedimentary structures that tell us about their previously unknown origins, and geologists in the 1960s who studied their structures finally divined their secrets. These geologists were also among the first scientists in North America to apply what they observed in modern environments to ancient sedimentary deposits, and, just like the ecologists, they did this in Georgia.

Starting in 1964 and through 1967, geologists John H. Hoyt, Robert J. Weimer, and V.J. "Jim" Henry documented the sediments and sedimentary structures in these lengthy linear sand ridges and then linked them to what they observed in modern environments on the Georgia coast.[11] Through this integrated approach, they successfully showed that the sand ridges of southeastern Georgia were actually former dunes and beaches of ancient barrier islands. These sand ridges, which today are barely discernible rises on a mostly flat coastal plain, are more or less parallel to the present-day shoreline, with each ridge representing a sea-level high during the past few million years on the Georgia coastal plain. Geologists then applied a mix of Native American and colonial names to each of the ancient barrier-island systems: Wicomico, Penholoway, Talbot, Pamlico, Princess Anne, and Silver Bluff.[12] Of these the Wicomico is both the geologically oldest and the geographically most inland, whereas the Silver Bluff is the youngest, while also nearly coinciding in both height and position with the modern shoreline. Indeed, the southernmost barrier islands of the Georgia coast—Cumberland, Jekyll, St. Simons, Sapelo, St. Catherines, and Ossabaw—are composite islands, made of both Silver Bluff sediments left more than forty thousand years ago and new sediments pasted onto them in just the past few thousand years.[13] So you might think of these islands as old stone houses whose owners just recently added a new seaside porch with a fresh coat of paint.

How did these geologists figure out that a bunch of sand hills were actually lost barrier islands? After all, sand by itself is not a reliable criterion for interpreting an ancient environment. If you automatically associate sand with beach, think instead of how many sand dunes form far from any ocean,

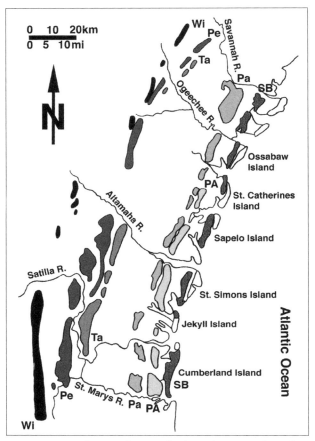

FIGURE 4. Map of ancient barrier islands on the Lower Coastal Plain of Georgia, showing past shorelines: Wicomico (Wi), Penholoway (Pe), Talbot (Ta), Pamlico (Pa), Princess Anne (PA), and Silver Bluff (SB) Islands. Map from Martin, 2013, 39.

accumulating in deserts or along the shorelines of freshwater lakes. Hence these geologists had to find clues that more precisely identified their origins.

How these geologists correctly interpreted the ridges is a testament to the power of ichnology and trace fossils. How did they do it? They first observed modern burrows on Georgia shorelines, then made a connection between the distinctive forms of these burrows and nearly identical fossil burrows in the ancient deposits. This was a perfect example of a principle deserving a nine-syllable name: "uniformitarianism." When applying uniformitarianism, geologists assume that certain geologic and biological processes and the results of those processes have remained more or less uniform over spans of geologic time. For example, because we can visit a Georgia beach today and see sand shaped into ripples by flowing water and wind, we can reasonably surmise that one-billion-year-old sandstones with ripples on their tops were likewise made by flowing water or wind. More recently, geologists have begun using the less syllabic but still somewhat mysterious term "actualism," which nonetheless conveys the same perspective.[14] In short, when we observe what is actually happening today, we can use those observations, often summarized neatly by geologists in the pithy phrase "the present is the key to the past," to guide us. And that is exactly what those Georgia geologists did in the 1960s.

Does uniformitarianism (or actualism) have its limits? Yes indeed. For instance, in your daily activities today, have you seen any gigantic, fifty-ton sauropod dinosaurs walking around? (If so, please let us know right away.) Atmospheric oxygen concentrations have not always stayed the same either, reaching levels around 30 percent at some points in earth history, which is significantly more than the current 21 percent.[15] These elevated amounts of oxygen allowed for the evolution of drone-sized dragonflies with seventy-centimeter (twenty-seven-inch) wingspans and encouraged ferocious forest fires that ignited with the slightest of sparks.[16] Differences in the oceans, atmosphere, biota, and many other fluctuating conditions in the 4.6 billion-year history of the earth all remind us that uniformitarianism is at best a broken funhouse mirror, one that supplies a fractured and distorted view of what may or may not have happened in the past. As a result, geologists are reasonably skeptical of making one-on-one analogies between what we learn from the present and whether their keys will unlock the mysteries of the past: the deeper the time, the less certain the comparison.

Fortunately for coastal geologists studying the Georgia coast and coastal plain, the temporal distance between past and modern processes affecting these places is short, and ghost shrimp (*Biffarius biformis* and *Callichirus major*) have been there for all of it. Ghost shrimp are distantly related to the free-swimming crustaceans many people enjoy in various culinary preparations but will never adorn a cocktail, no matter how trendy the restaurant. Ghost shrimp are so called because they are almost never seen, living in deeply plumbed burrows far below our feet on any given beach. But once these shrimp make themselves apparent, they are nearly transparent, neatly reaffirming their nicknames.

Ghost shrimp are decapods (ten feet), arthropods that include crayfish, lobsters, crabs, hermit crabs, and swimming shrimp and more narrowly belong to the evolutionarily related group (clade) Axiidea.[17] Biologists more specifically refer to ghost shrimp as callianassids, which paleontologists have documented going back to at least 150 million years ago, into the Jurassic period.[18] One of the behavioral traits common to all callianassids today, though, is their ability to burrow, and impressively so. In fact, burrows very similar to those made by modern callianassid shrimp are in rocks dating to the Carboniferous and Permian periods, from about 300 to 250 million years ago.[19] I have personally seen these fossil burrows in Permian strata of Brazil and Australia and was awed to think that these trace fossils may represent homes ancestral to those constructed by modern callianassids.

You may be relieved to know that you do not have to be a geologist to find modern ghost-shrimp burrows on nearly every sandy Georgia beach, but you will have to think like a volcanologist. Just go to that beach at low tide and look for what look like tiny shield volcanoes.[20] A burrow occupied by a ghost shrimp will complete that allusion by erupting water from its narrow, pencil-lead-thin aperture at its top. But this water often includes the occupant's mud-filled fecal pellets through a narrow aperture. Despite their decidedly unappetizing origin, these small, dark cylinders bear an uncanny resemblance to chocolate sprinkles one might see on cupcakes or doughnuts. (No matter how tempted you might be to incorporate these into recipes, please just leave them be.) Just below the beach surface, the apertures plumb straight downward for about twenty centimeters (eight inches), then widen considerably, encouraging visual allusions that stray from volcanoes and connect better with wine bottles. This widening of burrow shafts accommodates their ghost-shrimp occupants. These decapods move up and down the shaft to irrigate their burrow by pumping out their unwanted feces (understandable, that) and to circulate oxygenated water into the burrow. Balls of muddy sand reinforce the burrow walls like bricks in a house, stuck together by all-natural shrimp spit, and the burrow interior is lined with a smooth wall of packed mud and sand.

Amazingly, burrow shafts of ghost shrimp descend vertically far below the beach, as much as 2–3 meters (6.5–10 feet) deep.[21] Once at such depths, they turn horizontal, oblique, and vertical, and tunnels intersect, branch, and otherwise render a complex tangle of piping, perhaps reminding baby

FIGURE 5. Ghost-shrimp burrows on Sapelo Island, modern and ancient. *Left*, vertical shaft of modern ghost-shrimp burrow eroding out of a beach on Cabretta Beach, Sapelo Island. *Right*, vertical shaft of a fossil ghost-shrimp burrow eroding out of an outcrop in what is now maritime forest on Sapelo Island, but we know used to be a shoreline because of this trace fossil. Scale in centimeters.

boomers of playground jungle gyms they enjoyed as children in a prelitigation world. What happens down there in such adjoining ghost-shrimp burrow complexes, away from prying human eyes? Only the ghost shrimp know. Regardless, these burrows are restricted to the lower areas of beaches covered by tides (the intertidal zone) and sandy areas just offshore below the tide (the subtidal zone). Hence if you found similar burrows in the geologic record, you could reasonably infer where you are with respect to an ancient shoreline.

I think you now know where this is going, and how geologists figured out what geologic processes were responsible for the sand ridges on the Georgia coastal plain. Before these geologists did fieldwork in those areas, they may have already suspected that the sand hills of the coastal plain—now covered by pine forests, grasslands, agricultural fields, or strip malls—were associated with former shorelines. So with such a hypothesis in mind, they were probably thrilled to find fossil burrows that matched modern ghost-shrimp burrows they had seen on the Georgia coast but preserved in ancient sand deposits. They also found these fossil burrows in Pleistocene-age deposits on Sapelo Island, which helped them to figure out where the shoreline was located about forty thousand years ago with respect to the present-day one.[22]

In a shining example of how geology is a real science that involves testing hypotheses through repeatable results, I have found these same fossil burrows in Pleistocene deposits on the west side of Sapelo Island. Even better, I found them while leading a geology field trip with my students. They had seen modern ghost-shrimp burrows just the day before on the present shoreline, yet the fossil ghost-shrimp burrows were separated from that shoreline by several kilometers and in a maritime forest. So my students learned for themselves the same lesson imparted on geologists from the 1960s, that the Georgia barrier islands were made of both Pleistocene and modern sediments. In this respect, the amalgams of two shorelines and the coinciding of the ancient and modern in one place make the Georgia barrier islands unlike most other barrier islands in the world. This also means the Georgia barrier islands are near-perfect places for ecologists and geologists alike to discern universal principles for their respective sciences.

Geology and ecology combined further later in the 1960s and 1970s, when paleontologists who also were well trained in biology began looking at how ghost shrimp, ghost crabs (*Ocypode quadrata*), marine worms, and many other animals change coastal sediments—and, by extension, entire ecosystems—through their behavior. Were these scientists considered geologists, biologists, or ecologists? They were actually greater than the sum of their parts: they were ichnologists. And what they found through their studies of modern traces on the Georgia barrier islands made the islands even more scientifically famous, aiding in their recognition worldwide as among the best places for scientists who compare modern traces with those of the past.

3

Georgia Salt Marshes,
the Places with the Traces

My appreciation for the importance of the Georgia coast in the history of
the natural sciences didn't really start until I moved to Athens, Georgia, in
1985. Even so, the full realization took a while longer, maybe not hitting me
until more than two decades later. I attributed this delay to my reading clin-
ical descriptions of coastal environments in scientific journal articles rather
than having in-depth discussions with people who studied these environments (better) or direct experiences in those environments with those same
people (best). Once these interactions happened, my appreciation grew. But
nowhere did the significance of the Georgia coast become more obvious
than when I considered its salt marshes.

The dawn of this personal enlightenment began at the end of my master's degree program in geology at Miami University in 1985. As all gradu-
ates of Miami enjoy reminding those not in the know, this university is not
in Florida. Because of a more famous Florida city of the same name, it is
commonly described as "Miami of Ohio." The school directly confronts this
confusion by selling T-shirts in the campus bookstore that proudly (and ac-
curately) claim, "Miami Was a University When Florida Belonged to Spain."
(Miami University was founded in 1809, whereas Florida did not join the
United States until 1845.)[1] So it took my leaving the mostly comfortable (if
somewhat icy) confines of Oxford, Ohio, for PhD study in the so-called Deep
South of Georgia before stepping onto the shores of that state.

The Georgia coast has long captured the attention of scientists interested
in its biological and geologic systems and how these two realms overlap. As
mentioned earlier, ecologists there in the 1950s began investigating the exchange of energy and matter between the plants and animals of the Georgia
barrier islands.[2] In particular, they were drawn to the salt marshes, most

of which are between the mainland and upland parts of barrier-island eco-systems. Why study salt marshes in Georgia and not somewhere else? And how do the traces of plants and animals in these marshes, such as root dis-turbances, scrapings, burrows, and feces, actually play a major role in the functioning of these ecosystems?

First, knowing a bit more about salt marshes might help with under-standing how ecologists used them to develop basic principles of ecology. Georgia salt marshes are flat, extensive coastal "prairies" dominated in their lowermost parts by a tall marine-adapted grass called "smooth cordgrass" (*Spartina alterniflora*).[3] If Georgia needs any ecological bragging points, it should lead with these cordgrass-ruled salt marshes, which are among the most productive of all ecosystems, besting or equaling tropical rain forests in this respect. Moreover, despite the distribution of Georgia salt marshes over only about 160 kilometers (100 miles) of coastline, they still represent about one-third of all salt marshes in the eastern United States by area.[4]

How did this happen? Such an unusual concentration of salt marshes along the relatively small Georgia coastline is a result of several factors. One is its semitropical climate, which rarely dips below freezing during the win-ter, allowing marsh plants and animals to thrive and actively participate in their ecosystems nearly year-round. Another is the relatively high tidal range of the Georgia coast, which is about 2.5–3 meters (8–10 feet); in con-trast, tidal ranges along most of the U.S. Atlantic coast are well below this.[5] This range means enormous amounts of organic material—living and non-living—are cycled in and out of marshes by great volumes of moving water.

A third reason, and perhaps the most important, is what people did not do to the marshes, which was develop them in ways that would have completely altered their original ecological characters. Take a look at the barrier islands of New Jersey as examples of what could have happened in Georgia. (Just don't look too long: it's not pretty.) Salt marshes that are not drained, filled in, paved over, or otherwise irreparably altered can be studied and appreci-ated for what they are, not what we suppose.

The original scientists interested in the Georgia salt marshes, such as Eugene "Gene" Odum, Lawrence "Larry" Pomeroy, and John Teal at the University of Georgia (Athens) Marine Institute on Sapelo Island, were as-tonished by the amount of organic matter produced in these marshes.[6] They were especially taken aback by what was happening in the lower parts, ap-propriately called "low marsh." Much of this flux is controlled by tides and just five species of organisms a casual visitor can easily see any given day in

FIGURE 6. Tracemakers of the Georgia salt marshes, making them special. *Top*, close-up look at eastern oysters surrounded by smooth cordgrass, working together as ecological allies to produce, trap, and accumulate mud in a salt marsh on Sapelo Island. *Bottom*, mud fiddler crabs and their many traces on a Sapelo salt marsh surface, including feeding pellets, scrapings, and burrows.

these marshes: smooth cordgrass; ribbed mussel (*Geukensia demissa*); eastern oyster (*Crassostrea virginica*); marsh periwinkles (*Littoraria irrorata*); and mud fiddler crabs (*Uca pugnax*).

Just to oversimplify matters, but to assure that you get the big picture, the flow of matter and energy goes something like this. Smooth cordgrass is the primary producer of organic material in the salt marshes, converting sunlight into food for it and—as it turns out—lots of other organisms. This is a relatively easy task for these plants because they are powered by free-range and all-natural Georgia sunlight much of the year. Smooth cordgrass also has extensive and complicated root and stem systems, which help to hold marsh muds in place when marshes are flooded or flushed by twice-daily tides.[7] These roots and stems also locally change the chemistry of the surrounding mud and otherwise leave visible traces of their deeply penetrating networks, some noticeable long after the plants have died and decayed.

Before any mud can be held in place by *Spartina* roots, though, it must be produced and deposited. What causes this to happen? If you said "clay- and silt-sized particles settling from quiet and still water," you need to review and reflect on what I just said about the tidal range. Sure, I understand why one might think a salt marsh is quiet and still, especially when gazing over an expanse of smooth cordgrass while reading Sidney Lanier's poem "The Marshes of Glynn" and sipping a mint julep.[8] But the enormous volumes of water coursing through its creeks and other marsh surfaces during flood-and-ebb tides ensure that any sedimentary particle smaller than $\frac{1}{256}$ millimeters wide will stay suspended for as long as it is in water and for as long as the earth has a moon.

This incessant suspension is partly because the water is constantly moving but also because the clay minerals composing clay-sized particles are platy. These sediments thus act like tiny Frisbees, with their thin, flat profiles keeping them aloft in the mildest of flows.[9] A glance, quick or otherwise, at the opaque, café au lait hue of a tidal creek in a marsh at low tide, high tide, or in between will confirm that the water is full of clays. In short, they will not settle.

Then how does mud in a salt marsh get out of the water and onto the bottom to help make a marsh? I will let you in on a little secret, one that even many scientists do not know: most mud in a salt marsh is deposited as sand. Yes, you read that right: sand. Do you need another stunning fact? Animals turn it into sand. What happens is that ribbed mussels, oysters, and similar suspension-feeding animals suck in the water that is made cloudy by suspended clays. These animals then filter organic goodies from this water, such as dead or living one-celled algae, or detritus from *Spartina*, or other formerly living materials that might prove nutritious. Whatever is not useful, however, is excreted. Among the roughage ingested by suspension feeders are—you guessed it by now—clays. Ribbed mussels take this a step further by packaging the clays with mucus, like a biological form of shrink-wrapping.[10] Voilà, these packages are sand sized. The mucus eventually degrades, but the deed is done, with mud making it to marsh bottoms.

Ribbed mussels and oysters further assist cordgrass roots by stabilizing marsh surfaces and banks of tidal creeks, which prevents mud from washing away with each ebb tide. Both the cordgrass and oysters also baffle and otherwise slow down the water flow, causing mud, fecal or otherwise, to get deposited. In short, a Georgia salt marsh with its thick deposits of beautifully dark, rich, gooey mud, much of which consists of the traces of mussels and oysters, would cease to exist without these bivalves and smooth cordgrass and would become more like an open lagoon. Meanwhile, marsh periwinkles constantly move up and down the stalks and leaves of the smooth cordgrass, grazing on algae growing on the cordgrass. This activity—done by millions of small snails—damages the plants, tearing off small bits that fall onto the marsh surface.[11]

Fungi and bacteria further break down this "gentle rain from heaven" of cordgrass debris once it reaches the marsh surface. There mud fiddler crabs consume this stuff, along with any algae that might be growing on marsh surfaces. Their scrapings and discarded balls of processed sediment are everywhere to be seen on the upper parts of marsh surfaces.[12] These balls add to the sediment load of a marsh by packaging and redistributing the mud. Furthermore, like their molluscan neighbors, fiddler crabs poop. These end products add yet more mud to the sediment load, meaning fiddlers and other species of crabs living in the salt marshes are also donating enriched organic material. Fiddler crabs have an additional ecological role in salt marshes by digging millions of burrows. This activity churns and aerates the uppermost part of the marsh mud, much like how earthworms mix soils in a forest or field.[13]

So just to review: a typical Georgia salt marsh is composed almost entirely of mud deposited as feces from ribbed mussels and oysters, which are traces. *Spartina* roots and stems course through and bind the mud, also making traces. The *Spartina* leaves bear the grazing traces of marsh periwinkles. The marsh surfaces are scraped and burrowed by millions of mud fiddler crabs, which are also traces. All of this means that visitors can't go to a Georgia salt marsh and say, "I can't see any traces," unless they are closing their eyes or otherwise sight deprived, because the entire salt marsh is composed of traces that are the products of plant and animal behavior. Furthermore, these traces actually control the ecology of the salt marshes. This is why I often refer to Georgia salt marshes as examples of "ichnological landscapes," places that are the sum of all traces.

This trace-centered concept, then, better prepares us for thinking of salt marshes and other Georgia coastal environments as places where geologists can learn much from the ecological interactions of their communities with their physical environments. And with that mental linkage made, these scientists and others accordingly gain a better understanding of how organisms can leave scripts both big and small in the geologic record. Some of these scripts last and some disappear, but those that persist always change.

4

Rooted in Time

As a paleontologist and geologist, time is always on my mind. In my experience most nongeologically inclined people assume these temporal musings of mine always involve millions or billions of years. Such expanses better connect with the so-called deep time so beloved by geoscientists and used to shock others accustomed to pondering much shorter intervals, like the life of a mayfly, a wait in the drive-through of a fast-food restaurant, or the length of a cat-themed video. As a result, the stories geoscientists tell about the history of a place often begin long before humans were in the picture and are told in chronological order, like flipping (or swiping) pages in a book from page one to the end.

The most honest of geoscientists, though, will also admit that entire chapters or other lengthy sections of their stories are missing, huge gaps that can be filled only by looking for those chapters in books elsewhere. This means they also test whether their "book" of time is missing a sentence or paragraph, or whether the "author" went back to rewrite a paragraph. Because of these Earth-as-editor realities, most paleontologists and geologists also try to interpret briefer lengths of time that range from a few minutes to hours to years.

One of the better tools for diagnosing these temporal snippets are trace fossils, such as tracks or burrows, which often provide snapshots of individual behaviors that happened in a blink of geologic time. Such precision is also helpful when remembering that these candid portraits were not staged in museum displays or reenacted by computer graphics but made by living things in the context of their original ecosystems. When cutting away at the historical record, ichnology thus becomes more like a scalpel and less like a chainsaw. Then, when combined with other geologic data, these vestiges of life allow our imaginations to become time machines, helping us to create vivid, evidence-based explanations that are later honed through further scrutiny and reimagining. Simply put, ichnology helps us bridge the gap between the recent, the old, and the very old.

For instance, when I look at a limestone layer in which its sediments were laid down ninety-five million years ago during the Cretaceous period, and I see thousands of small, U-shaped burrows in that limestone, I do not imagine the rock it is today. Instead, I visualize it as a soft mud at the bottom of a warm, clear, shallow lagoon, with little crustaceans digging into the mud and using their burrows as homes for suspension-feeding lifestyles. Furthermore, I think of the burrows representing just a few generations of burrowers that inhabited the lagoon for only a few years—maybe a decade—until it ceased being a lagoon. This happened after the sea level dropped and an overlying layer of sand or mud buried the lagoonal sediments, all of which much later turned into stone.

This assessment of a thin slice of a past environment and its biota might seem impressive in itself, but other trace fossils can also unravel even more complicated sequences of short time intervals, including its gaps. For instance, if I also found large carnivorous dinosaur tracks on top of those same burrows, and the burrows were obviously compressed by that one-ton animal, I might be tempted to say the dinosaur strolled into the shallow water of that lagoon and unknowingly crushed those poor little crustaceans as they cowered in their burrows. Yet if I look more carefully at the limestone, I will also see the dinosaur tracks barely compressed the surface. This tidbit of information tells me the mud was firm, not soft, as a weighty dinosaur should have sunken much deeper. This conclusion also implies that the former lagoon was no longer underwater but likely had been buried; its sediments were compressed and dehydrated and then later exposed. This sequence of events in turn implies that although the sediment had not yet hardened, the burrowing crustaceans were no longer alive. Then maybe only a few years passed before their former homes saw the light of day again and were summarily squashed by a Cretaceous version of Godzilla.

All of the preceding scenario is not some hypothetical exercise in time travel but is based on a real example in Dinosaur Valley State Park near Glen Rose, Texas. This park and the surrounding area are world famous for their abundant and high-quality dinosaur tracks in limestones from ninety-five million years ago, including the just-described limestone bed.[1] This layer not only bears thousands of small U-shaped burrows—telling of the former presence of suspension-feeding crustaceans in a shallow-marine lagoon—but also preserves the trackway of a large theropod dinosaur that stepped on a few of those burrows.

How did I learn about these rocks and their secrets? I was invited to join a research team from Indiana-Purdue University there in the summer of 2012, and one of my assignments was to study this layer and its burrows. The more specific goals of this little study were twofold. One was to discern how the burrows related to the original environment that produced the limestone. The second was to understand what happened between the time the crustaceans were happily burrowing and feeding in their little lagoon until a dinosaur strolled through their neighborhood.

Despite my location in the exotic land of east-central Texas, at some point during the week while I examined this limestone and other strata like it, I got Georgia on my mind, and more specifically the Georgia coast. As I stared at these former softgrounds that later became firmgrounds ninety-five million years ago, I recalled seeing analogous (but much younger) deposits on the Georgia coast that also preserve such tiny, within-a-human-lifetime gaps in time. These Georgia coast deposits show where former salt marshes thrived, died, and were buried, only to later become exposed and reborn as different environments. I also recalled how the evidence of past environments merged with those of another suite of biota while these sediments were still sediments. This has happened before, and it will happen again.

But before we jump from Texas to Georgia and from the Cretaceous period to modern times, please indulge my offering a brief review of how geoscientists consider fossilization and otherwise discern the passage of bodies and life traces into the geologic record. Can we actually watch fossils in progress, as the remains of the once living put themselves into history through either bodily remains or their traces of life? The answer is, yes, we can. And even if we cannot necessarily carry the experiment out to completion by traveling millions of years into the future, we can still gain a few helpful insights on how the fossils we see in a sedimentary rock got there.

Despite pitiful lamentations about the "incompleteness" of the fossil record, fossils are actually quite common. Geologists bolster this truism whenever they include trace fossils in a fossil checklist while examining any given outcrop of sedimentary rocks formed in the past 550 million years or so. Now, please forgive me as I briefly criticize a few (and it really is only a few) of my geologic colleagues, but I must. The sadly unenlightened individuals who warrant my ire are those who still hold onto the long-outdated concept that bodily remains from the geologic past—such as leaves, shells, and bones—are "real" fossils. In contrast, they choose to diminish or ignore the tracks, burrows, trails, and other indirect evidence of former lives in sedimentary rocks. Hence a baby kitten dies (metaphorically) whenever I visit an outcrop filled with trace fossils that other geologists described as "lacking fossils." (For those of you who dislike infant felines, this metaphor may vary. But I will ask you regardless to imagine something precious winking out of existence, such as doughnuts, spiced lattes, or miniature golf.) In short, what those fossil-negating people actually meant by "lacking fossils" was "no body fossils." Usually these unrecognized trace fossils are invertebrate burrows, but tracks and other trace fossils may also reveal themselves to those who look for them.

Indeed, the expectation of finding fossils of either type in a sedimentary rock is not just optimistic; it is realistic. This means that on occasions when geologists find a sedimentary rock layer devoid of either body or trace fossils, we scratch our heads and ask why the original environment had no visible signs of life. Because living things bigger than microbes have been part of our planet for about a billion years, bodies and traces—whether rare, abundant, or somewhere in between—are the norm, and their absence an anomaly.

How do the former body parts of plants or animals, or traces of their behaviors, become preserved as fossils in the first place? This question and others related to it are addressed by the science of taphonomy. First coined by Russian paleontologist Ivan Yefremov in 1940, this word stems from the Greek roots *taphos* (burial) and *nomos* (law).[2] The "burial" part makes sense, as one of the most effective ways to preserve evidence of a former life is to bury it. That way it is not subjected to the elements: wind, rain, decay, scavengers, monster-truck rallies, and more. The "law" part of this term, however, is a little unconventional, as most scientists might have simply added *ology* to the prefix, as in *taphology*. Nonetheless, Yefremov's choice of suffix implies that he expected the natural processes causing fossil preservation to be orderly and predictable and that they followed certain rules. Of course, paleontologists studied fossilization before this branch of their discipline was defined by a new word.[3] Also, forensic scientists since Yefremov's time have overlapped their studies with paleontologists, sometimes resulting in their cooperation as they excitedly share gruesome insights on dead bodies. This does not necessarily mean paleontologists are solving crimes, nor are we fighting (or committing) crimes, but the division between paleontologists and criminologists is not so far.

A pithier way to describe taphonomy, and one I use with my students and anyone else hearing about it for the first time, is to describe it as "the study of everything that happens to an organism after it dies." This description accounts for how dead bodies that enter the fossilization lottery typically need to get buried. Once under a pile of sediment, they undergo some sort of alteration, whether biological, chemical, or physical.[4] But before that burial and alteration, these bodies also might bloat, float, disarticulate, and otherwise go through a complicated postdeath history rivaling the epic journey depicted in the 1989 film *Weekend at Bernie's*. For example, because we know all dinosaurs lived in landward environments, we also know that dinosaur bones in marine deposits indicate their bodies must have floated out to sea. In other words, just because something was dead doesn't mean it stopped moving.

Before this exploration of taphonomy becomes far too lengthy and stultifying, let's return to the Georgia coast and examine how taphonomic principles can be applied there. The most superb case in point—and the one that entered my thoughts while I was deep in the heart of Texas—is a relict marsh on Cabretta Island. This island is part of the Sapelo Island complex, divided from the main part of Sapelo by a tidal creek on its west side and a broader tidal channel—Blackbeard Creek—to its east. The Cabretta relict marsh is what remains of a salt marsh from several hundred years ago, one that was buried then but exposed on a modern-day sandy beach by a combination of a shifting shorelines and a rising sea since at least the 1970s.[5]

The most remarkable aspect of the Cabretta relict marsh, though, is the convincing clues it gives reflecting its original identity. Its evocative traits have drawn me back to see it dozens of times, with more than half of those visits as field trips with students in tow. My primary goal in taking students there is to give them a better appreciation for how a sedimentary deposit transforms from a once-living ecosystem to inert rock, while also retaining compelling evidence of its formerly teeming life. Similar relict marshes are on St. Catherines, Jekyll, and other Georgia coast islands, but when it comes to teaching about taphonomy in the field, I prefer the Cabretta one.[6]

As mentioned in the previous chapter, modern salt marshes on the Georgia coast have five key components that make them among the most productive of all ecosystems: smooth cordgrass, marsh periwinkles, mud fiddler crabs, oysters, and mussels. So if a Georgia salt marsh were to be buried quickly—say, by a hurricane or another storm that dumps a thick

FIGURE 7. A relict marsh on Cabretta Island (Sapelo). *Top*, overall view, with eroded stems of smooth cordgrass sticking out of the sand, firm mud below, and me for scale. The photo is from 2004, so the scale is now a little wider. Photo by Ruth Schowalter. *Bottom*, closer view of relict marsh on Sapelo Island, showing remains of smooth cordgrass hundreds of years old, a cross-section of its muddy (but compacted) sediments, and quartz sand deposited on top by tides, waves, and wind.

layer of sand on it—a taphonomist would ask this simple question: which of those five would get preserved? The Cabretta relict marsh partially answers that, showing us bodies and traces of some plants and animals. Granted, the clues in this example are "only" a few hundred years old, so they are incipient and merely auditioning for the role of fossils. Yet these parts and marks are on their way to becoming part of the geologic record, giving us a glimpse of the early fossilization process well before its completion in an uncertain and distant future.

Perhaps the most striking difference between the relict salt marsh and a modern one is easily demonstrated by stepping onto the relict one. If I tried doing this in a modern salt marsh, I would at least get muddy, or at worst get stuck and lose a shoe or two. In contrast, when I repeat the same experiment with a relict marsh, I can bounce up and down on its surface while singing Billy Idol's "Dancing with Myself." This difference in substrate consistency is all about water. For instance, if you say a salt marsh is full of it, you might be talking about bivalve feces, but you also might be referring to its water content. A typical marsh mud contains massive amounts of water, which is held between clay- and silt-sized particles, adding great volume to that mud while also bestowing its mucky properties. But if you buried this same mud under a thick pile of sand, the weight of the overlying sediment would squeeze much of the water out of it, reducing the volume of the deposit to a third or fourth of its original thickness.[7] This dewatering then makes the mud firm enough to invite my students to stand on it with me as I teach them about relict marshes. This "classroom floor" also introduces my students to the geologic concept of diagenesis, the chemical processes affecting sediment as it changes from soft to firm to rock.

The next obvious distinction between the Cabretta relict marsh and modern marshes revolves around what is left of the *Spartina* there. The tall green or golden stalks and leaves of smooth cordgrass, adorned by millions of bead-like marsh periwinkles, are absent from the relict marsh. Instead, only straw- to ochre-colored stubs testify of *Spartina* stems formerly connected to these leaves of grass. Extensive roots and rhizomes (underground stems) below what used to be the marsh surface further verify former populations of *Spartina*. These roots and stems are astoundingly complicated, traveling in fractal-like patterns throughout the muddy sediment, and quite deep, penetrating to nearly a meter below the surface.[8] While I was a graduate student, I remember reading how marshes are held together by *Spartina* root systems and thinking, "Okay, that makes sense." But once I saw cross-sections of those root systems in the Cabretta relict marsh, this previously nebulous concept became more solid.

Once in a while I also find old marsh periwinkle shells scattered on the surface of the relict marsh, which presumably represent a few of the many that once lived on the now-absent stalks and leaves. Periwinkle shells are made of calcium carbonate, which dissolve and otherwise break down quickly in slightly acidic waters. Once exposed, they might not last long. But the real reason for why these shells tend to disappear quickly is because they are spirited away by modern hermit crabs. When hermit crabs encounter these periwinkle shells on the relict marsh surface, they say the hermit-crab equivalent of "Hey, free shells!" and happily trot away with their finds, not caring that their newly adopted homes are actually hundreds of years old. Hermit crabs have been documented elsewhere using fossil shells if necessary, so it is not surprising that they would usurp slightly less old ones.[9]

But what about mud fiddler crabs? So far I have not seen bodily remains of these essential cogs from the original, fully functioning marsh ecosystem. This absence is not surprising, though, as fiddler-crab exoskeletons are composed of chitin (the same stuff making up our fingernails), which breaks down much faster than calcium-carbonate molluscan shells. Nonetheless, all is not lost, as old fiddler-crab burrows are abundantly evident on the relict-marsh surface as perfectly round, open holes. Those holes connect to lengthy vertical burrows that form J or Y shapes, visible in the same cross-sections with cordgrass roots.[10] Amazingly, these burrows are sometimes accompanied by new burrows made by modern fiddler crabs, as well as those of modern marine clams that drill into these now-firm surfaces. To preserve the burrows as trace fossils for future geologists, all this shoreline needs to do is fill their deeper parts with windblown sand.

Modern, living ribbed mussels are easy to see in Georgia salt marshes if one is willing to sacrifice footwear to the warm embrace of soft, dark, sulfurous mud. A boardwalk or other walkways above a marsh at low tide allow for less messy viewing of these bivalves, but if these are not available, my students often have to take my word for it that mussels are indeed in the marsh. Mussels attach themselves to marsh surfaces with fine threads called byssae, orient themselves vertically to suspension feed, and often clump together.[11] What convinces my students that I am not just making this up? Clumps of vertically oriented mussels on the relict-marsh surface help, making me look almost knowledgeable. A question I often pose to my

FIGURE 8. Old mud fiddler–crab burrows in the relict marsh. *Left*, longitudinal view (surface above and going down) of fiddler-crab burrows and smooth-cordgrass root traces. Fill the deeper parts of these burrows with sand, and they're more likely to get preserved as trace fossils. Scale on right side: fifteen centimeters (six inches) long. *Right*, eroded relict marsh surface with circular cross-sections of old fiddler-crab burrows, now being filled with modern beach sand. Scale in centimeters.

students then is, Now that these shells have been exhumed and exposed, how long will they last on the surface? The short answer is, not long. These, like the periwinkle shells, are made of calcium carbonate, and hermit crabs have no use for them, hence most of these subfossil mussels will soon break apart, dissolve, or otherwise vanish.

In my experience oyster shells are less often encountered in the Cabretta relict marsh than *Spartina* nubs and roots, fiddler-crab burrows, and ribbed mussels but are not as rare as periwinkles. Given the right exposures, though, we see old oysters on some visits. These lucky sightings are not of lone shells, looking as if they were discarded at a local coastal seafood feast. Instead, they, like the mussels, are clumped. Where mussels normally make groups of a dozen or fewer, oysters congregate by the hundreds or thousands, attaching to one another to form impressive masses of tightly cemented shells. Oyster-shell clusters in the Cabretta relict marsh typically

FIGURE 9. Cluster of old ribbed mussels around nubs of smooth cordgrass on the relict marsh surface; sandal (size 8½ men's) for scale.

form curving lines, outlining the banks of long-gone tidal creeks that meandered through the original marsh. Aerial drones have helped us better define these former creeks, which stand out as sinuous, ghostly interruptions in yellowish expanses of dead and denuded *Spartina*.

Given all this body and trace evidence of this once-living marsh on Cabretta Island, its story seems like a straightforward linear narrative that starts with "Once upon a time" and ends with "and they died happily ever after." But throw modern plants into the story, and this story abruptly introduces visitors from the future who write entirely new passages on top of the previous text. This mixing of old and new happens when modern plants decide to grow in the same medium occupied by previous but now departed plants.

Plants have no morals or ethics, and they certainly don't care whether or not they mess up our neat little linear-time stories. In fact, plants are so good at altering histories that they can lead you to doubt all geoscientists when they try to chronicle the events of an ancient land-based ecosystem.

Such misgivings are rendered because modern plant roots can occupy old sediments, rocks, or even bones.[12] Also, if their roots penetrate deep enough into these sediments, they may leave both remnants of their original tissues and root traces alongside those of much older (and long-dead) plants. Geologically fresh root traces may even mix with older animal trace fossils, conjuring the illusion of a contemporaneous community that all lived together in ecological harmony. Only a careful examination of the sediment, and which traces cut across which, can unravel tales of lives greatly separated by time.

This realization came to me twice during separate trips to the Cabretta relict marsh. The first plant-induced revelation was during a March 2012 visit, when I saw that part of this ancient ecosystem hosted a healthy stand of sea-oxeye daisy (*Borrichia frutescens*). This observation surprised me for two reasons: one, that any modern plant could grow in this firm mud, and, two, that this particular plant could. Sea-oxeye daisy is a shrub that grows to about a meter high, but usually less. It has light-green and lance-like fleshy leaves and yellow flowers that form spiky hard fruits. Up until then I had seen it only along higher areas around salt marshes and near coastal dunes. Later, after reading about sea-oxeye daisy and its incredible adaptability and hardiness, I was not so stunned that it moved into this old neighborhood. Sea-oxeye daisies not only tolerate saline conditions but also can grow in a variety of substrates, from muddy to sandy, from alkaline to acidic, and from sediment to rock.[13] So twice-daily inundation by tides and hard-packed marsh mud were not deal breakers for this plant. It normally reproduces by flowers, but it can also send in the clones by using bits of itself to take root or just use its roots. For the latter it simply spreads its rhizomes far and wide, making one individual plant on a relict marsh surface seem like many. This situation was likely what I saw, but the upshot is that the daisy had added the effects of its living roots and stems to the assemblage of ancient bodies and traces.

The next plant-related revelation came five years later, and it was even more stunning. A few colleagues of mine from Emory University and I were on a visit to Sapelo Island in 2017 to film short educational videos for a website we were producing, the *Georgia Coast Atlas*. As we approached the Cabretta relict marsh, I was shocked to see that part of it held a waving field of green. There on the north end of the former marsh was a thriving crop of young green *Spartina* growing in and among the eroded and yellowed

Spartina nubs of yore. Somehow this fresh crop colonized a newly exposed and eroded part of the old marsh, its living stalks popping up from around the dead. Were these zombie plants, having been reanimated by either the mutagenic chemicals of nearby paper mills or leaking radiation from passing nuclear submarines? Sadly, no. Yet this unexpected botanical oddity still brought me pleasure because of its peculiarity. Here were modern plants of exactly the same species occupying the same real estate as their ancestors, adding their parts and traces to this future fossil assemblage. Photos and videos taken that same day by an aerial drone (piloted by one of my colleagues) confirmed the extent of this cordgrass colonization.

As a result of these two modern plants in the Cabretta relict marsh messing with my taphonomic mind, I have since revised my appraisal of any sedimentary rock containing root trace fossils, and I never assume they represent the same living plant community. Instead, I think about descendants and ancestors mixing their effects in one fluid mess. The takeaway message I learned from the Cabretta relict marsh and others like it on the Georgia barrier islands is that the geologic record does not always tell a straightforward narrative. The concept that old environments can influence and blend with those of the present and that gaps in our narratives must be minded collectively cultivate an awareness of how environments can change both through time and within time.

Part II

Shells and Carapaces

5

Coquina Clams, Listening to and Riding the Waves

Cumberland Island is a scenic place, with long, sandy beaches; lush maritime forests; and superb salt marshes. As the southernmost of the Georgia barrier islands, it is also the farthest from the major metropolitan city of Atlanta and a good haul even from the coastal city of Savannah, Georgia. Also, unlike Jekyll Island—its barrier-island neighbor to the north—it does not have a bridge connecting it to the mainland. It is reachable only through twice-daily ferries or private boats. This isolation lends to a sense of exclusivity, while also making a visit there feel well earned.

Despite these geographic barriers, Cumberland still receives tens of thousands of visitors each year. Its popularity is partly because it is a U.S. National Seashore managed by the U.S. National Park Service, which enhances its visibility. The main attraction of Cumberland for many tourists, though, is its wild horses (discussed in great detail in another chapter). Tragically, Cumberland is not attracting hordes of tourists because of its tiny, burrowing, attentively listening, surfing, and migrating clams. Still, I hold out hope that that this criminal neglect will vanish once more people know about these bivalves and their extraordinary abilities.

I first learned about the Cumberland clams when reading about them in the first decade of the 2000s, and finally got to see them in action in February 2012. The occasion for this happy circumstance was my coleading a class field trip on Cumberland Island with three colleagues from Emory University. I had been to the island several times before, but all the students and two of my colleagues had not, casting this already-special place in a more exciting light for them. As is typical with any of my field trips to a Georgia barrier island, I noticed new phenomena while there. This was yet another example of how field trips with students ideally inspire wide-eyed wonder in their instructors too, demonstrating that discoveries can be made and shared while neatly nullifying jaded cynicism. Of the various finds that day, the intellectual highlight for me was the bivalve-related revelations provided by coquina clams (*Donax variabilis*) on and in the beaches of the southeast side of the island.

Our group first noticed the clams as a death assemblage just above the uppermost part of the surf zone on the beach. Their empty shells—some evident as single valves and others as pairs still hinged together—were deposited by waves following a high tide, then moved slightly afterward by the wind. These finely ribbed and shiny shells also justified why they bear the species name of *Donax variabilis*, as they display a gorgeous assortment of colors: yellow, orange, beige, blue, pink, mauve, and other schemes that surely would elicit gasps from interior decorators.[1]

Then, just a bit lower on the beach and in the freshly scoured intertidal zone, we noticed clusters of small sandy bumps. Underneath these bumps were living coquina clams that had buried themselves. You may remember that this is the same behavioral strategy used by another small clam, the dwarf surf clam, and the motivations for both species are similar. This burying helps the clams avoid drying out between tides, while also deterring predation by ravenous shorebirds. This tactic is better than nothing, but, as we learned earlier, these clams are still easy targets for shorebirds intent on acquiring fresh bivalve snacks. Imagine a person ducking under a blanket to avoid being eaten by a lion and how well that might work. (And, please, just think about it and do not test it on your next safari.)

Nonetheless, instead of simply writing off all coquina clams as inept burrowers that deserve to die at the beaks of their avian overlords, we should look well below the surface, and I mean the sand surface. So I accordingly directed our students to examine the spaces between bivalve bumps for pinprick holes in the sand. Once the students saw them, I then asked if they were single or paired (the holes, that is). The correct answer—pairs—accordingly led to the next rhetorical question: why? These, I explained, are the traces of siphons, extendable fleshy tubes connecting the more deeply buried coquina clams with the beach surface while keeping their delectable bodies out of sight.[2] These clams are thus much more likely to escape molluscan-munching birds, while also keeping themselves moist in saturated beach sands until the next high tide.

Coquina clams are accomplished burrowers, a necessary adaptation for nearly any small animal living in the high-energy surf of a Georgia beach. In the event of a wave breaking on a beach that washes away the uppermost

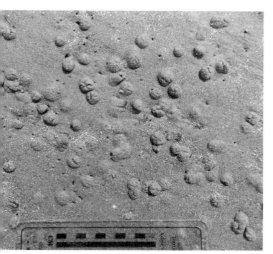

FIGURE 10. Coquina clams living and dead on Cumberland Island. *Top*, resting traces (or are they escape traces?) of coquina clams in the upper intertidal zone of a beach. These clams are buried just underneath each bump of sand, but others are much deeper and safer. See those little holes in between the clams? Those are siphon traces from more deeply buried clam compatriots. *Bottom*, no-longer-alive coquina clams, hence not making traces, some with both valves intact or apart, some partially covered by windblown sand. Scale in both photos in centimeters.

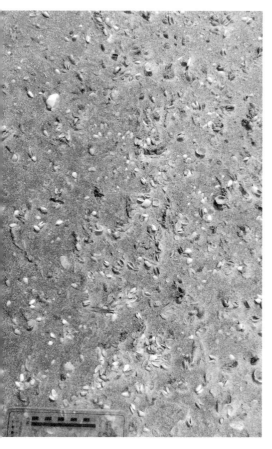

layer of sand and exposes a coquina clam, it will open its valves only enough to stick out its foot, which it then vibrates rapidly.[3] This movement loosens the wet sand underneath, and the clam's smooth, streamlined shell does the rest of the job, allowing it to glide into its self-made local pit of quicksand and vanish from the surface. I have watched coquina clams do this by the hundreds and in unison: when revealed by a wave, they collectively vanish before the next wave arrives.

Once under the sand and temporarily safe from further erosion, this clam orients itself vertically and projects its siphons upward, making paired holes visible at the sand surface. The complete burrow made by a coquina clam is thus Y-shaped, with the clam body making the lower, thicker part of the Y and the two siphons leaving two thinner traces above.[4] When we observed their traces on that Cumberland Island beach, I suspect the more deeply buried clams had the benefit of wetter sand when the tide dropped, whereas the ones hiding under mere caps of sand did the best they could with less wet sand as the tide continued to recede.

How do we apply this bivalve lore to the fossil record? Paleontologists have found similar small, Y-shaped burrows in fine-grained sedimentary rocks. Some of these are interpreted as the works of suspension-feeding bivalves, although polychaete worms or small arthropods are possible makers too.[5] In other instances bivalves burrowed down, oriented themselves horizontally, and stopped in the sediment below. This pausing sometimes left impressions of the clams' lower halves that were later filled by overlying sediments, which produced almond-shaped trace fossils.[6] As a result, paleontologists can reliably identify a former presence of bivalves in rocks that might not have retained any of their shells, and coquina clams help us better apply this actualism to ancient examples.

What else is significant about coquina clams? For one, their shells were abundant enough in the recent geologic past to have formed the framework for a loosely cemented limestone common in Florida, appropriately called "coquina limestone." This is a rock encountered by nearly anyone who has enjoyed (or suffered through) an introductory geology lab exercise on sedimentary rock identification. Students of mine often compare these rock samples to popcorn balls or similar sugar-cemented treats, although I've also noticed that no matter how long I kept the students in lab, no one has been tempted to eat them.

Anyway, how these thin little shells later became rock is an interesting story in itself. During the latter part of the Pleistocene epoch—tens of thousands of years ago—coquina clams thrived in Florida and Georgia beaches, especially during the warmer times between glaciations. When generations of these clams died and were buried, massive accumulations of their calcium-carbonate shells were compressed together and partly dissolved by groundwater. This groundwater then later reprecipitated calcium carbonate as cement, gluing the shells into a cohesive and mappable rock unit called the Anastasia Formation.[7]

This seemingly fragile rock composed of billions of coquina clams that burrowed into coastal sands to protect themselves was also later used by humans for their own protection. Soon after the Spanish invaded North America during the fifteenth and sixteenth centuries and began constructing missions and forts along the coasts of Georgia and Florida, they sought durable materials that would withstand cannon attacks from their European rivals, the British and French. On Anastasia Island in Florida, the most obvious source for such materials was the local coquina-limestone bedrock. In the late seventeenth century, the Spanish quarried this limestone to construct Castillo de San Marcos, a fort intended to protect the then-new settlement of St. Augustine.[8]

Did this clam-supported defensive tactic work? Yes, it did. Unlike more solidly compacted rocks that would have fragmented on impact, the porous limestone actually absorbed the impact of cannonballs fired at it, with the experiments unwittingly conducted by British naval forces in 1702 and 1740.[9] This successful application of limestone deterrence delayed the British acquisition of Florida until 1763. As a result, one could say this coquina limestone—and by default its clams—had a major effect on the recent human history of North America.

Colonialism and military history aside, I saved the best tidbits of information about coquina clams for last, which I expect most people do not know, and ones that makes these clams, like, totally cool. These little bivalves respond to sound and migrate seasonally. Yes, that's right: these clams have their own form of "listening," and they can move en masse once the seasons change on the Georgia coast.

Despite persistent rumors that clams have legs, they definitely do not have ears. Thus, they do not hear in the sense we do but instead respond to low-frequency vibrations caused by waves striking the shore.[10] Once they detect these vibrations, they react. First, they pop out of the sand and "jump" up into the water, each clam propelled by its muscular foot. Once out of the sand and into the water, they bodysurf on the wave, allowing the momentum to carry them landward. When the wave breaks on the shore and the clams are dumped by it, they quickly rebury themselves in the sand. Like aficionados of heavy metal or punk rock, louder is better, as higher-decibel waves cause more clams to jump up out of the sand and into the water, neatly rendering an aquatic imitation of a molluscan mosh pit.

Also, coquina clams—like caribou (*Rangifer tarandus*), wildebeest (*Connochaetes* spp.), and Arctic terns (*Sterna paradisaea*)—migrate. Granted, coquina clams do not cover the vast distances of those other migrating animals but simply move up and down the slope of a beach with the changes of the seasons. Through a winning combination of wave surfing and burrowing, they move from the lower intertidal zone—where they live during the spring, summer, and fall—to the upper part of the beach, which becomes their winter homes.[11] Hence coquina clams do the opposite of snowbird retirees who abandon higher ground for coastal environments during the winter by instead moving up as temperatures go down.

In short, I hope that all this pondering over a few shells, bumps, and holes on a Cumberland Island beach has helped lend an appreciation for the small wonders on any given Georgia barrier island. Who knows what little discovery the next field trip to a Georgia barrier island will bring, or whether some facsimile of what is seen there might also be preserved in the fossil record? As the old saying goes, time will tell, whether that time is in the present or the geologic past.

6

Ghost Crabs and
Their Ghostly Traces

The Atlantic ghost crab of the Georgia barrier islands, common to beaches from New England to Brazil, is among my favorite tracemakers anywhere, anytime. My ichnological admiration for ghost crabs in general—comprising more than twenty species with a worldwide distribution—stems from the great variety of behaviors they record in beach and dune sands, leaving behind fascinating tales of what they were doing while no one was watching.[1] Their light and speckled bodies also help them blend with local beach sands, effectively disguising them when they remain still. This invisibility, compounded by their nocturnal inclinations, thus lend easily to apparitional allusions. So I thought it only appropriate that with a common name including "ghost" as part of their description, these crabs deserve a story about some of the enigmatic clues they leave of their unseen behaviors, reinforcing their reputations as the poltergeists of the beaches.

On the dawn of June 22, 2004, on Sapelo Island, my wife, Ruth, and I were presented with one of the most intriguing ghost-crab mysteries I had seen related to their traces. We were walking along the freshly scoured surfaces of Nannygoat Beach on the south end of the island, scanning it for traces. High tide only a few hours before had cleansed the beach of traces, erasing the page of the previous day's scribbles and allowing its residents to start anew. Low-angle rays of early morning sunlight optimally contrasted any freshly made animal signs on the beach, another reason why we were there then. We also went to the beach with our minds open to anything novel.

Sure enough, within about fifteen minutes of stepping foot on the beach, Ruth paused and asked one of the simplest—yet most important—scientific questions: "What is this?" She pointed to a depression on the sandy surface in the freshly washed intertidal sands, and what I saw astonished me. It was a trace perfectly outlining the lower (ventral) half of a ghost crab, preserving

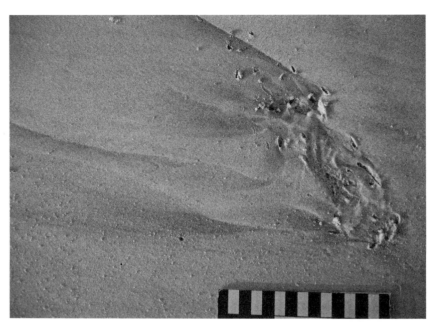

FIGURE 11. A ghost crab leaves a mysterious (and perfect) outline of its bottom half on a Sapelo Island beach, traces found just after dawn and high tide. Note the tracks leading away from it, showing a live crab made it. Scale in centimeters.

in detail impressions of all eight walking legs (pereiopods), the pointed ends of these legs (dactyli), its smaller claw (inferior cheliped), larger claw (superior cheliped), and its main, rectangular body.

This perfect imprint was exceptional in and of itself but was made downright odd by other traces. Connected to this "body" was a single set of ghost-crab tracks away from it; none moved toward it. Hence these traces were not made by the dead body of a crab as a death mask, preserved before it was plucked from the beach by a gull or wave. Instead, the tracks showed the crab was very much alive when it made its resting trace and walked away from that spot. But what happened just before then? For all we could tell, it looked as if the crab floated through the air, dropped straight down, made a perfect ten-point landing onto the beach, sat a while, and walked away.

Knowing that ghost crabs are capable of performing many tricks, but not aerial acrobatics, we wondered how the traces were made. Working as scientists, though, we also knew that single observations are only the start of good science. For this inquiry to progress any further, we had to see if our seemingly unusual observation could be repeated. So we walked farther south along the beach to test whether this ghost-crab impression with its single set of tracks was an isolated incident or if more such traces existed. We succeeded. With our search image and a goal in mind, we quickly found about a dozen more such imprints made by crabs of various sizes, showing an identical behavior. Even better, all were located just below the high-tide mark of the previous night, clearly having been made at about the same time. We now had not datum but data.

Time to think. These crabs must have walked to their resting places, but why did they not leave any tracks showing this? We soon figured those tracks were certainly made, just not preserved. Hence, like all other surface traces on the beach, they must have been erased during high tide. Yes, that was it. The crabs walked to the surf zone just after the high tide, sat down, waited long enough for the tide to drop a bit, and walked away.

Mystery solved? Well, not quite. This was an incomplete explanation, one with a big, unanswered question and one related to motivation. Why did the ghost crab walk to the surf and sit down? (With a prompt like that, feel free to create your own intertidal-crab versions of chicken-crossing-road punch lines.) Ghost crabs normally spend much of their time in deep, J- or Y-shaped burrows close to or in the dunes and above the high-tide mark.[2] They are most active at night, when they come out of their burrows to scavenge delicious dead bodies of various sea creatures dumped on the beach by waves and tides, prey on smaller invertebrates in the sand (like dwarf surf clams), or scrape up sandy surfaces to eat algae.[3] They also leave their burrows to seek mates, which might involve one crab enticing another to visit its burrow.

Yet none of the crabs that made these traces were scavenging, preying, or mating. Something else in the surf beckoned them, offering a life-sustaining reward that compelled them to risk becoming meals for night-stalking predatory birds. I searched my memory for what I had read previously about ghost crabs and their biological needs and finally realized what could have driven them to the surf in the middle of the night. They were thirsty.

You see, ghost crabs are living examples of so-called transitional animals that evolution deniers insist do not exist, because they possess a mix of adaptations to radically different environments. Ghost crabs are descended from fully marine crabs, meaning they still have gills that filter dissolved oxygen from marine waters. Yet they also have little lungs and can breathe air, enabling them to stay out of the water for hours.[4] Having both gills and lungs makes ghost crabs semiterrestrial, living in a world between the land

and ocean and depending on resources from both realms. They still live close to the sea for their food, reproduction (females lay their fertilized eggs in the ocean), and water, but their main livelihood is gained from the beach and dunes, where they live and eat.[5]

In this respect ghost-crab burrows in the upper parts of beaches and lower parts of dunes provide protection against predators but also keep the crabs hydrated. One of the functions of a ghost-crab burrow—which can be more than a meter (3.3 feet) deep—is to intersect the seawater table below.[6] That way, when a crab needs water, it crawls down the burrow to that saturated area and replenishes its precious bodily fluids. But they cannot stay down there as the tide rises. Unlike blue crabs (*Callinectes sapidus*), which are fully adapted to marine environments, if a ghost crab is immersed in seawater too long, it drowns. So like many intertidal crustaceans, including fiddler crabs, they move higher up to pockets of air trapped at the top of the burrow.[7]

Another clue helped better define the story. The night before our discovery of the ghost-crab resting traces and tracks, Nannygoat Beach experienced a spring tide related to a full moon. This higher-than-normal tide probably flooded many ghost-crab burrows and caused these crabs to abandon their homes. This, in turn, meant these crabs might have spent most of the night outside their burrows, which contributed to their dehydration. Yet they also had to wait for the high tide and its accompanying energetic waves to recede before risking a trip to the ocean edge. As soon as the tide turned and dropped and the waves became less vigorous, the crabs ran to the surf zone, settled down into the wet sand, and soaked up water through small openings where the legs connect to the main body. Spiky "hairs" (setae) on their legs help with this water uptake, drawing up moisture from the sand through capillary action, like putting a paper towel on a puddle of water.[8] Ghost crabs are amazingly efficient at pulling water out of sand. Their hunkering down onto a saturated sandy surface, with gentle waves breaking on top of them, must have been like the ghost-crab equivalent of drinking from a funnel, quenching their thirst in a most satisfying way. Meanwhile, waves washed away their tracks leading to these resting spots. They stayed in these places a while, long enough for the tide to drop and expose the sandy beach surface. Only then did they get up and walk away, fully rehydrated and refreshed, ready to look for more food, go back to their burrows, or dig new ones.

This is a detailed explanation, but one based entirely on traces and what little I knew then about ghost crabs from the scientific literature. How else to test it and see whether or not it was right? If you just said, "By directly

FIGURE 12. A ghost crab caught in the act, making a body outline and tracks. *Left*, a ghost crab that doesn't mind getting a soggy bottom. This one sprinted from the dunes to the surf, stopped abruptly, and sat down for a spell. *Right*, that same ghost crab ready for its close-up, so thirsty that it didn't mind my taking its picture. Scale in centimeters.

observing this interpreted behavior in a ghost crab," then that's a bingo. A little more than a month later, on July 30, 2004, I actually witnessed a crab making these traces, again on Nannygoat Beach. On that visit I was Ruthless and by myself and looking for more ghost-crab traces after a high tide. While walking along the beach this time, though, I saw a small, wraithlike movement out of the corner of my eye. It was a beautiful, full-sized adult ghost crab, flat out running in full daylight and heading straight from the dunes to the surf zone.

I stood back and watched it reach the surf, where it promptly sat down and stilled. As I walked toward it, I took photos from afar, expecting it to bolt at any moment. Instead, I was surprised to see it remain where it sat, even as its eyestalks rotated to look warily at me. Surprised, I figured this one must have been thirsty enough to stand its ground and risk being eaten or stomped. My resulting photos showed just how close I got to it, with direct overhead shots that showed every bump on its carapace and all setae on its legs. I was also thrilled to see this crab in exactly the same position depicted by the traces Ruth and I had seen the month before. It was almost as if science has predictive power that can be tested with further observations that later lead to refutation or confirmation.

Although scientists are certainly not always right, if we practice good science, we sometimes hit the nail on the head, or the crab on the sand, or, well, never mind. With the "resting trace equals rehydration" hypothesis now supported by both traces and direct observation, I wrote the results into a formal, peer-reviewed paper that I submitted to a scientific journal. Unexpectedly, such traces had never been documented for ghost crabs, especially from the perspective of a paleontologist. In my paper, which was accepted and published in 2006, I pointed out how this behavior would explain similar-looking trace fossils in the geologic record or the preservation of fossil-crab bodies frozen in the same position by death.[9] I speculated that perhaps at least a few fossil crabs represented times when they reached the surf too late and were buried by wave-borne sands. The geologic significance of such trace fossils would be their value in pointing exactly to where a wave may have washed across an ancient shore, millions of years ago. Geologists really like this kind of precision and become indebted to ichnologists who give them such tools they can easily apply in the field.

Since 2004 I have seen these resting traces on the beaches of nearly every Georgia barrier island, in the Bahamas, and other places where ghost crabs dwell. As another example of how ichnology and paleontology are predictive sciences, trace fossils echoing this behavior in ghost crabs or their ancestors were reported by paleontologists elsewhere, who used exactly the same search images provided by the Georgia coast ghost crabs.[10] Given this ichnological lesson from the ghost-crab traces of the Georgia coast, my hope is that these marvelous animals have become just a bit less "ghostly" and much more real and alive in our imagination.

7

Ghost Shrimp Whisperer

When you hear the word "shrimp," you probably picture bacchanalian displays of these formerly swimming arthropods on ice in grocery stores or on the plates of restaurants throughout the world, consumed voraciously by their terrestrial admirers. Another type of shrimp, mantis shrimp (such as *Squilla empusa*), which fortunately are eaten much less by people, have also received some attention lately. This admiration is well deserved, and, given their rainbow-colored bodies and deadly forearms that can punch and kill small fish, they are among the most gorgeous and terrifying of marine invertebrates.[1] But other marine crustaceans bear the common name "shrimp" that neither grace seafood buffets nor are awesome predators, yet still merit our adoration, documentary films, and epic songs worthy of performing on Eurovision. Yes, you guessed it: I'm talking about ghost shrimp.

Why ghost shrimp? Because they can burrow like nobody's business. Take a typical ghost shrimp in the Bahamas or the Caribbean, such as *Glypterus acanthochirus*. This crustacean is only about 10 centimeters (4 inches) long, but if it lives for eight years and burrows continuously through that time, it will have processed a cubic meter of sediment.[2] Individual ghost-shrimp burrows also can go down as deep as 5 meters (16 feet).[3] When scaled up, these feats of burrowing are comparable to a human shoveling more than a cubic kilometer of dirt, or a vertical shaft about 100 meters (330 feet) deep, but done without a shovel, backhoes, augers, drilling rigs, or other tools. Vertical shafts of these burrows also connect with extensive branching tunnels, creating complicated networks in the sediments below the low-tide mark. Now multiply the industriousness of a single ghost shrimp by millions, and we are talking about enormous volumes of sediment processed in their respective shallow-water environments. Ghost shrimp are thus like the ants of the shoreline, only not as organized: no queens, workers, soldiers, or other divisions of labor, just lots of individual shrimp burrowing, eating, mating, and defecating.

As decapods, ghost shrimp share a common ancestor with crabs, lobsters, crayfish, and other shrimp, all of which have four pairs of walking legs and one pair of claws. Mantis "shrimp," on the other hand (or claw), are crustaceans but not even decapods; they belong to an evolutionarily linked group (clade) of crustaceans called Stomatopoda.[4] Marine biologists and ichnologists also know ghost shrimp as thalassinidean shrimp or, more specifically, callianassid shrimp, belonging to the clade Callianassidae within Axiidea (mentioned in chapter 2).[5] These shrimp burrow through sand and mud using their front two claws and carry or otherwise move sediment with their other legs. As mentioned before, ghost shrimp are also well known for depositing much of the mud on Georgia beaches as elegantly packaged little cylindrical fecal pellets outside of their volcano-like burrow tops. (Who could forget those alluring chocolate sprinkles?)

Geologists and paleontologists love ghost shrimp too, because of how their burrows are so numerous, fossilize so easily, and are sensitive indicators of shorelines. As we also learned previously, callianassid burrows both modern and ancient helped geologists in the 1960s more easily identify and map abandoned barrier islands on the lower Georgia coastal plain.[6] Since the 1960s geologists and paleontologists worldwide have identified and applied trace fossils made by similar decapods in rocks spanning more than 250 million years.[7]

I could prattle on about ghost shrimp and their ichnological incredibleness for the rest of the book but will kindly spare you that grueling scenario. I will instead get to the point by bragging about one of my best traits: humility. Just when I thought I had learned nearly everything there was to know about ghost-shrimp ichnology, one shrimp decided I needed my eyes opened to traces I had never before seen and to learn something new about them. This mind-widening experience happened while I was teaching undergraduate students from my barrier-islands class on Jekyll Island in mid-March 2013. We were on its northernmost sandy beach—often nicknamed "Driftwood Beach" because of the many dead trees foundered in the surf there—when we encountered an odd little series of tracks. The trackway was connected to a funnel on one end and to a shallow, horizontal excavation capped by sand on the other end. This suite of traces turned out to be the home burrow, tracks, and temporary hiding place of a ghost shrimp. How did I know a ghost shrimp made these traces? Perhaps because they were associated with a live ghost shrimp, but that's beside the point.

Anyway, before I made the big reveal with my students, we looked at the trackway together for anatomical clues that would identify the tracemaker. The trackway was small, only about a centimeter (less than ½-inch wide), but lengthy, with an irregular and meandering course that traveled more than a meter (3.3 feet). It had a symmetrical series of pinpoint impressions on either side that looked like fine stitching. This pattern definitely pointed toward an arthropod as its maker and not a snail, clam, polychaete worm, or other legless animal. In between those tracks was a barely noticeable groove, left by some middle part of the arthropod: not legs but something else, like a tail. The "funnel" at one end was a collapsed crater in the sand, only about 4 centimeters (1.5 inches) wide, but with a deeper inner pit in its center. At the other end of the trackway was the horizontal tunnel, about 15 centimeters (6 inches) long and slightly less than 2.5 centimeters (1 inch) wide, a little longer and wider than my index finger. The tunnel was denoted by a roof of sand that reminded me of covered bridges I had seen during my Indiana childhood but differed considerably in both size and form. It was a lumpy, bilobed, zipper-like pattern that roughly defined a series of chevrons. I was sure it also had tracks underneath, and I was likewise pleased to realize the chevron pattern on top was a result of legs underneath it that rhythmically pushed back sand with each forward movement of the arthropod's body.

My students, caught in this enigma with me, watched as I squatted to look more closely at this tunnel. It was open in its middle at one end, marking where the cohesive, wet sand roof had collapsed. I could not see if its maker was below this little skylight but did not need to once I looked at the opposite end of the tunnel. There was its tail, protruding from its sandy cover as if playing hide-and-seek but having chosen a carpet too short for full concealment. The tail was translucent gray rimmed with light brown and divided into three overlapping fans, similar to what I had seen on the tails of certain invertebrates in grocery stores and restaurants. That is when I knew it was a shrimp of some sort, yet when I carefully scooped the sand from underneath it with my fingers and held it up for all to see, I was as astonished as my students.

It was definitely a ghost shrimp. Although I could play a marine biologist on TV, I am not one, but I still hazarded to identify it as a Georgia ghost shrimp. This species and the Carolinian ghost shrimp are very common, living underneath Georgia beaches and in shallow, offshore sediments by

FIGURE 13. Sometimes you look down on a Jekyll Island beach at low tide, and you see something odd. Notice the collapsed top of a ghost-shrimp burrow is in the lower left, but it's connected to a little trackway, which ends in a shallow horizontal burrow, which holds the maker of all three traces. Lots of other ghost-shrimp-burrow tops are in the upper part of the photo. Scale in centimeters.

FIGURE 14. A tracemaker reveals itself. *Left,* close-up of the shallow horizontal burrow, a short tunnel with a roof of sand. Check out the slightly collapsed roof on the right and that its tail is sticking out on the left, like hiding under a too-short blanket. *Right,* ta-da! Here's who was hiding under that sand, a Georgia ghost shrimp, just before I put it back in the ocean so it could get back to burrowing.

the millions.[8] But what truly surprised me about this particular one and its traces was their rarity. As I said to my students then, after more than fifteen years of fieldwork on the Georgia coast, I had never seen anything like this sequence of traces, let alone leading to a live ghost shrimp. To have the tracemaker right there was like the period at the end of a sentence in its story, providing a most satisfying conclusion.

Nonetheless, what was truly fascinating about the story told by these traces was its beginning (collapsed funnel) and middle (trackway), speaking of individual crisis. For this shrimp to behave so unusually, something must have threatened its life. Ghost shrimp almost never see the light of day and prefer to stay deep in their burrows, well away from the prying eyes and beaks of shorebirds, fish, and other predators. Consequently, they remain largely invisible to humans, partly justifying the "ghost" part of their common name. This means something very bad must have happened to this one in its burrow, forcing it to abandon its refuge and expose itself so vulnerably, like a fire forcing people out of fortified underground bunkers while knowing tyrannosaurs lurk just outside. Damned if you do, damned if you don't, but something in this ghost shrimp's evolutionary programming made it take the path of the lesser damned.

What happened? Did a subsurface predator find its way into the burrow and chase it out? Was it a chemical cue of some sort, like oxygen-poor water flooding the bottom of its burrow? Was it competition from another ghost shrimp, evicting it from its home? Was it a mate that decided it had enough of sharing this burrow and needed some alone time or took up with a comelier shrimp? I didn't know, but this missing piece in the puzzle added intrigue.

The story of the ghost shrimp told by the traces on this Jekyll Island beach might also be viewed as a hero's journey in a three-act play. In the first act it was living in its safe, comfortable burrow, a perfectly normal existence that ensured it was living an ordinary life. But, suddenly, something terrible happened while it was there, giving it good reason to leave its burrow and venture onto the alien landscape of the surface world. For more than several minutes, it wandered in this dangerous place, experiencing its open air, scents, and noises, all while feeling the earth tremble from the footsteps of giants roaming this land. At some point in its journey, exhausted by its travails but wishing for a longer life, it did something it had never before done. Instead of digging a deep vertical shaft, it dug a shallow tunnel. There it stayed hidden, until a few curious (but gentle) giants stopped to investigate this little oddity in their landscape, one of whom uncovered the ghost shrimp. Although at first discomfited at its discovery, it relaxed and accepted whatever fate awaited. The almost-wise older giant who held it in his hand then talked admiringly about it with the much younger giants. For several minutes they shared an appreciation for this visitor to their surface world. They then photographed it, thanked it for teaching them more about it, and placed it back in the ocean, where it burrowed happily ever after. Unless it died, that is. The end.

With this lesson learned from the story of a single ghost shrimp, my students and I gained new perspectives of its species and how trying circumstances can lead to unusual behaviors. The little story further told by its life traces also again demonstrated the rewards of taking time to study the imprints of a small, individual animal and how such close examination can kick-start our imagination. Will any of us—or anyone else, for that matter—see something like these traces again, and with a ghost shrimp at its end? I hope so, if for no other reason than to know that others may be likewise inspired by such marvelous revelations.

8

Why Horseshoe Crabs Are
So Much Cooler Than Mermaids

Are mermaids real? If you answered no, then please continue reading. If you answered yes, then put down this book and go back to 2012 and 2013. That was when a cable TV network decided that broadcasting a teasing, hour-long pseudodocumentary on these mythical marine animals (2012), followed by a sequel (2013), had priority over producing and airing real documentaries on real marine animals.[1] To add insult and salt-laden ocean water to injury, both programs got great ratings. As one might imagine, marine biologists were livid and took action. With a minimum of regard for decorum, these so-called trained and educated scientists wrote outraged tweets and indignant blog posts and otherwise took umbrage in a most uncivil way.

So rather than waste any more time and energy addressing such intellect-eroding excuses for entertainment that have happened then and since, I will instead highlight one real and extant marine animal that seldom fails to fill me with awe and reverence, wiping out any need to worship mythical animals. This animal's lineage is more ancient than those of dinosaurs, reptiles, or amphibians, with its oldest fossils dating from about 450 million years ago.[2] It is also the largest living marine invertebrate animal you are likely to see on beaches of the U.S. East and Gulf Coasts. Like Aquaman (but far more erogenous), this animal lives in the ocean but can also walk on land for hours. At the start of summer each year, if you see it crawling around on a beach, it's because of sex. From May through June this species begins its massive orgies on beaches from the Gulf of Mexico to Maine.[3] During these debauches, much larger females are approached by many smaller males that shove and jostle one another on their way to fertilizing. Pictures of this gamete-laden frenzy somehow make it past prudish censors of social-media sites every year, titillating prurient invertebrate enthusiasts everywhere and filling them with cockle-warming glee. One might say that this animal deserves its own planet.

As you already know from reading the title of this chapter, I am talking about horseshoe crabs. More properly known as limulids by genuine marine biologists and paleontologists, these ultracool, überhip, but also totally retro critters are misleading in their common name. Horseshoe crabs are actually more closely related to spiders than true crabs, but their common name is so, well, common that scientists just sigh and begrudgingly go along with it for the sake of public dialogue.

Modern limulids are represented by four species, three of which are in Asia. The Atlantic horseshoe crab (*Limulus polyphemus*) is the largest species of horseshoe crab in the world, and those in Georgia are the largest of that species.[4] Such grand sizes may be a function of the Georgia Bight, an extensive offshore shelf that affords more food and habitat than other areas where wild horseshoe crabs roam. How big are they? I've seen some as long as seventy centimeters (twenty-seven inches)—tail included—and forty centimeters (sixteen inches) wide, big enough to scare even the most overfed of domestic cats. They grow to these sizes after hatching as little limulids not much bigger than the period on this sentence, an astonishing increase in mass if they make it to adulthood (which most do not).

People who see these animals for the first time cannot help but notice each limulid has a prominent head shield; this part is its prosoma. On the prosoma are two obvious compound eyes on either side, but if you look closer toward its front, you will also see small, simple eyes behind each compound one and three simple eyes toward their front. Horseshoe crabs even have eyes underneath their prosomas that they use for "looking" along the sea bottom and light sensors on their tails. (If you ever wonder whether or not a limulid is looking back at you, it is, and in multiple ways.) Its prosoma is streamlined for moving through the water and plowing into mud and sand but also protects important parts underneath, which include its legs and mouth. Their first pair of legs is modified for shoving food into their mouths (mostly little clams), and the remaining five pairs are for walking. But the first pair of walking legs differs in males and females, as the males have little grappling hooks for latching onto females.[5] Otherwise, limulid legs have little pinchers on their ends that aid in walking, and, unlike ghost crabs, fiddler crabs, and other true crabs, these parts pose no danger whatsoever to any part of human anatomy. The spiny part attached by a hinge behind the prosoma is a limulid's opisthosoma; this is what protects its book gills, so called because they lay on top of one another like pages. The book gills are also where females carry their eggs.

Because of these and many other limulid qualities, I went on a pilgrimage to Delaware in May 2018 to watch them mate and learn more about the traces they leave of their wanton activities. (This is also a great time to see huge flocks of migratory shorebirds, such as red knots [*Calidris canutus*], which time their migrations to feast on the bountiful abundance of limulid eggs.)[6] Fortunately, other folks have written entire books about horseshoe crabs and heaps of popular and scientific articles, so I will not needlessly duplicate what they have done so well. Instead, in the following I focus on my main interest in these animals—their traces—and especially regale you with tales of the traces they make with their tails.

Limulid enthusiasts have no doubt noted that I neglected one of the most important of all horseshoe crab parts, the spiky projections behind their opisthosomas. These "spikes" are their telsons. On the basis of many traces on the Georgia coast, both my direct observations and those of other people, I can tell that the main function of a telson is to help a horseshoe crab get back on its feet after it's been knocked onto its back. Whenever a limulid is upside-down, it immediately starts using its telson as a lever to right itself into a less vulnerable position. Without a telson, an upside-down horseshoe crab is stuck; its legs run furiously but to no avail. With a telson, however, it can put the pointy end into the sand or mud underneath its body and push itself up from a surface. This action gives a limulid a fighting chance to walk toward the ocean and otherwise get back to where it once belonged. Such actions work best if the overturned limulid turns itself to its right or left, as they are longer than wide. Although horseshoe crabs are wonders of nature, they are not doing back flips or somersaults. Incidentally, this is also why you should never, ever pick up a limulid by its tail: if it breaks, you just assured its eventual slow, agonizing death with most of its eyes buried in the sand.

I had already known all the preceding information about horseshoe crabs for quite a while. After all, up until 2015 I had coauthored a book chapter about juvenile limulid traces and their close resemblance to trilobite trace fossils; coauthored another article on the history of limulid-trace studies (which go back to the 1930s); and devoted a lengthy section of a chapter in another book, waxing eloquently on limulids as tracemakers.[7] So you could say I felt pretty cocky about what I knew about these animals as tracemakers—that is, until one horseshoe crab, much like that ghost shrimp on Jekyll

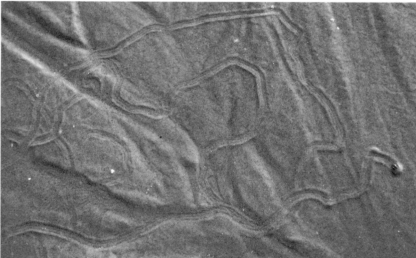

FIGURE 15. Atlantic horseshoe crabs and their kid-friendly traces on Sapelo Island. *Top*, a fine juvenile specimen of a horseshoe crab, walking on land and making tracks, and if this one lived long enough, it used those legs for walking on land and making baby horseshoe crabs too. *Bottom*, the circuitous trail of a baby horseshoe crab. You can estimate its body width from the width of the trail interior, and its body length was slightly more than that, meaning it was much smaller than my fingernail. See that central groove? That's from its tail, but if you want to impress your friends, call it a telson.

Island, showed me how much I still need to learn about them and what they can do.

The modesty-inspiring traces showed up in a photo on a Facebook page I followed, the St. Catherines Island Sea Turtle Conservation Program. On that page the program organizers then—Gale Bishop and Robert "Kelly" Vance—regularly added photo albums of sea-turtle traces (trackways, body pits, nests) and otherwise reported on other facets of natural history they observed on St. Catherines Island beaches. As a result, while I was marooned in the metro-Atlanta area, I could still live vicariously through their pictures. Granted, most of what they shared was already familiar to me, but they also liked to throw me ichnological stunners once in a while, such as the ones they posted one day in July 2015.

Kelly found these enigmatic traces while patrolling the beaches of St. Catherines Island for other traces, namely those of expectant mother sea turtles. Although they distracted briefly from his mission, I was grateful that he stopped to document these oddities, as they were new to me. They consisted of a number of holes in the beach sand that defined a nearly perfect circle. These holes were made by the telson of an adult female horseshoe crab that kept trying to right herself after landing on her back. Each puncture mark shows where she inserted the telson into the sand and then pushed herself up and to her side. As shown by the number of holes, direction of sand flung out of each hole, and little "commas" made by extraction of the telson, she tried to flip herself a minimum of sixteen times, all to her right. These separate actions culminated in a 360-degree clockwise rotation of her body. This series of holes surrounded a central depression with smaller drag marks, which is where the limulid's prosoma contacted the sand. To imagine the movement represented by these traces, think of a horseshoe crab doing a slow motion, step-by-step, break-dance backspin. I would suggest trying to embody this behavior by reenacting it while wearing a horseshoe-crab body suit, complete with telson attachment, over a few hours. Sadly, though, I suspect that even eBay does not have such items available for purchase. Yet.

Seeing the evidence for this limulid nevertheless persisting was worthy of inducing wows, but in my ichnologically influenced euphoria I also figured she finally succeeded in righting itself. After all, the trackway just to the left of the traces indicated she walked away from the scene of her gravitationally challenged situation. But then I realized there was no impact mark: a large female horseshoe crab flipping herself onto the sandy surface should have registered an outline of her body before walking. Instead, the

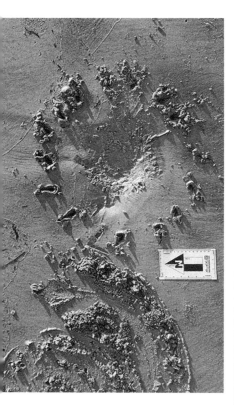
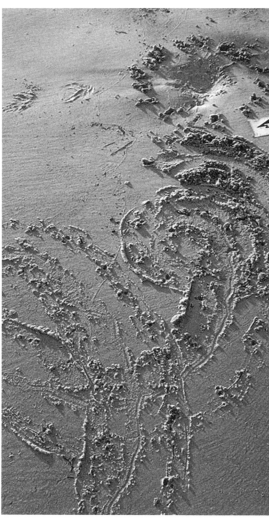

FIGURE 16. Traces of horseshoe-crab distress on St. Catherines Island. *Left*, circular pattern of pokes in the sand made by its telson trying to right itself, with the worn area in the center showing where its back rotated on the sand. *Center*, looping tracks spiral away from the place of distress, where she was saved by an intervening bipedal land-dweller. *Right*, almost there, with the horseshoe crab making her way to the sea after her ordeal. Photos by R. Kelly Vance.

place where she took her first steps showed no such impression, meaning she made a soft landing with only her legs and telson digging into the sand. What happened? Did she use mind over matter and levitate herself through telekinesis? Or was she gently picked up and placed on her feet by a merciful mermaid? (Or merman: let's make sure we're being inclusive when talking about made-up stuff.)

It turned out that Kelly was the deus ex machina that entered this limulid's drama by providing divine intervention just when she needed it. After I expressed my puzzlement to him about how this large arthropod finally turned over, he confessed to saving her, in which he lifted her and put her back on her feet. From there she promptly walked away in a series of tight spirals. The spiraling is something I had seen before in their tracks, a method limulids use to find the downslope direction that normally leads them to the low-tide mark and the comfort of a watery environment.

But it is not all sunshine and lollipops for many other horseshoe crabs, which still get stuck, telson levering aside. Given the tidal range on the Georgia coast (2.5–3 meters, or 8.2–9.8 feet), strong wave energy, and wide beaches, many big limulids that come in with the flood tide get knocked onto their backs and left behind. Although I was glad to see how these traces promised a happy ending for this limulid that so stubbornly tried to get upright, I also acknowledge how many of her horseshoe-crab compatriots did not make it back into the life-sustaining sea. This reality also serves as a reminder that storybook endings do not always happen in nature and what we wish to be true sometimes is not. One might think some sort of natural selection is taking place and that similar travails might have happened in the geologic past, affecting the evolution of its lineage.

In this instance I don't know whether this horseshoe crab made it back into the sea to live another day or not. Still, the lesson she left for us in the sand lives on, and I am now slightly more confident that if any limulids were stuck on their backs at any point in their 450-million-year history and made similar traces with their tails, and these marks were preserved as trace fossils, we just might recognize them for what they are. For that alone I am appreciative. Thank you, horseshoe crabs, for being real and for making traces on the Georgia coast, then and now. It is an honor to share this planet with you today in this very thin slice of your 450-million-year history.

9

Moon Snails and
Necklaces of Death

Picture yourself walking down a Georgia beach, perhaps with a loved one, or someone you want to love you, or someone you want to love, or not. As you stroll, a waning tide reveals the biological secrets of the sea for both of you, whether as traces or strewn body parts. Among the latter are near-perfect shells of clams and snails, their beauty enticing, imploring you to pick them up, hold them, and praise them. But much like your respective personalities, many of these shells bear a potential flaw: a perfectly round hole.

You look closer and see the hole is beveled, wider on the outer surface of the shell and narrower inside. In the clams it is normally placed near its umbo, closer to its original hinge than its edge. After seeing a good number of these pierced shells, you decide they are not blemished but made functional for personal adornment. This hole, you declare, is where we will insert a string to make a necklace. So you do not just stop with one shell, and both of you spend the next few hours selecting several dozen similarly sized ones that can be threaded together. This quest for perfect shells with their perfect holes bonds you to each other, and for years afterward, whenever one of you wears this necklace, you both smile in warm remembrance of how it brought you together.

The problem, of course, is that you are wearing a necklace of death—and not just of death but of horrible killing beyond the most fevered machinations of any science fiction, of suffocating embraces, dissolving tissues, and agonizingly slow erosion of defenses against an adversary that is never going to give you up, never going to let you down, never going to say good-bye, until you die. This intimate relationship involves not consummation but consumption, in which your corpse will be assimilated into the essence of your vanquisher so that it grows larger and stronger to kill another day, and another, and another. Such a necklace would have made Mary Shelley giggle and elicited a chortle from Edgar Allen Poe, and one surely lies in a desk drawer of Stephen King, who slips it over his head for inspiration before writing a particularly macabre passage.

The holes in these Georgia clams and snails are the traces of predatory moon snails (*Neverita duplicata*). Their common name is understandable once you look at this snail from above its uppermost whorls, as the outline of its shell describes a circle. But shell collectors also refer to them as "shark eyes" because of how the last few darkened whorls in the shell center may also resemble a pupil.[1] Unlike whelks, moon snails are flatter and fatter, wider than tall, and the largest are only about 10 centimeters (3.5 inches) wide, fitting in a child's hand. They have beautiful glossy exteriors bearing a swirl of finely defined colors—red-brown, purple, pink, yellow, and more—spiraling about their centers like tiny galaxies. If you are lucky enough to find a live one, you also may see its fleshy foot, which through its absorption of water can inflate to more than twice the volume of the shell.[2] Look quickly, though, as this foot typically retracts into the shell, "weeping" as it sheds water to accommodate the smaller space inside. This discharge happens just before an orange-tawny lid closes over the foot, flush against the shell opening. The lid is its operculum, protecting the animal's soft parts against prying eyes, beaks, or other means of assault from the outside world.

My PhD dissertation adviser, Robert "Bob" Frey, often referred to moon snails as the "lions of the tidal flat," a metaphor so evocative that I kept it alive whenever teaching others about them. Frey bestowed this nickname in recognition of these snails filling the niche of a top predator in the below-surface ecosystems of beaches. How do they hunt? Think of them less like sinewy denizens of a savannah and more like submarines, but in sand instead of water. They normally burrow in sediments offshore, but when stranded by a high tide onshore, they hide themselves in shallow (but sometimes lengthy) burrows.[3] Moon-snail burrows are distinctive as straight-to-curving ridges in wet sand exposed at low tide; sometimes these burrows end with a roundish plug of sand, under which you will find the live trace-maker. These plugs are only slightly more than the width of the moon snail that made it, giving you a prediction of its size.

Similar to whelks and other gastropods, moon snails burrow without legs or other limbs. They accomplish this by inflating and deflating their foot, which they use to intrude the sand beside or below them.[4] Through muscular contractions, as well as a series of anchoring and pulling motions, their smooth shells slip easily through saturated sands. Nonetheless, considering they are composed of mere calcite while moving through harder quartz sand, one might reasonably wonder how their shells stay so smooth. For instance, imagine dragging your body across sandpaper and its effects on your clothing and skin. In contrast, a burrowing moon snail makes its way through the sand unblemished by covering its shell with part of its foot and secreting mucus. This dual action provides a scratchproof protective coating while also lubricating the snail's passage, enabling it to keep slip-sliding away.

Moon snails use this stealthy approach to their advantage by attacking potential prey (usually clams or other snails) from the sides or below in the sand. Because they cannot see while buried, they use chemical cues to "sniff out" their potential quarries, a better chemistry through living that results in something dying. Water flowing between grains of sand delivers molecules exuded by moon snails' intended menu items, triggering the snails into action.[5] Sometimes this method is augmented by "sound," with a moon snail feeling vibrations in the sand and moving that way.[6] Once live food is detected, they may turn abruptly toward their targets, with gastropod decisions denoted by sharp turns in their burrows. Then the hunt is on. A snail's pace is just fine for these predators, as they only need to be faster than their meals.

When a moon snail catches up with either a sedentary or likewise moving molluscan in the sand, it extends part of its foot to engulf and immobilize. If its victim is a clam, the snail may manipulate it so that its proboscis "drill rig" is positioned near the clam's umbo.[7] Because this is where the clam's adductor muscles hold it shut, a successful attack there weakens a clam's defenses, making it more likely to submit and open up. Once a moon snail is in position, it commences drilling, using both chemical and physical tools. First the snail releases a small amount of acid and enzymes to weaken a spot on the shell. This weakened point is then rasped with a hard, breaking-and-entering tool inside the snail's mouth, its radula.[8] A moon-snail radula is like a tongue with teeth, which the snail sticks out and rotates against the shell. Although no one has yet developed a moon-snail cam to directly observe this process, we presume the radula turns right, left, or a combination of the two. But to produce a beveled hole, it must be applied at an angle to the shell surface, rather than parallel to it or straight down.

The time needed to complete drilling is more like hours, rather than minutes or days, and no doubt depends on the size of the prey versus the predator and thickness of its shell. Yet once the shell is breached, it's all over for the prey. An open hole allows the moon snail to insert its proboscis inside the shell with its radula on its end, and it eats its hapless meal from inside while it is still alive (albeit temporarily).[9] Once done the satiated moon snail departs, with only a barren shell bearing a distinctive hole left to state the moon-snail equivalent of, "Yeah, that was me. What are you going to do about it?"

In the preceding description I purposefully made the moon snail's ideal hunt sound as deadly and efficient as possible, possibly worthy of a B-grade sci-fi-horror movie adaptation of a gigantic moon snail amuck in a small, coastal Georgia community, leaving a mucus-lined trail between emulsified and shriveled victims punctured by beveled drill holes. Yet hunts sometimes fail. For about every twenty clams or snails I find with open holes testifying to clean kills, I find one with a partially made hole that did not penetrate the shell. This is evidence of a predation fail, where a moon snail found its prey, enveloped it, and began drilling, but its victim fought back and somehow escaped.[10] These traces of resistance and ineptitude often tell of a moon snail's ambitions woefully outstripping its abilities, in which it attacked a too-large clam or snail, like a house cat mistaking a Norway rat for a mouse. Moon snails may even attack empty and very-much-dead shells, which (adaptively speaking) is just downright dumb. Half holes also relate to calories gained versus energy expended, a balance sheet that favors those moon snails that go after appropriately sized and live prey. A moon snail that habitually bites off more than it can chew uses more energy than it gains, meaning it may not make it long enough to make it, and its genes will reach a literal dead end.

The little drill holes we see in shells on Georgia beaches accordingly hint at a much grander history, one of an evolutionary dance and coadaptation between predators and prey over hundreds of millions of years. In the fossil record drill holes in shells are nearly as old as shells themselves, with tiny holes in shelled organisms apparent from as long as 550 million years ago. While doing my master's thesis work in fossils from the Late Ordovician period (about 440 million years ago) in Ohio, I remember encountering beautifully sculpted holes in fossil brachiopods, similar to those I see in mollusks of the Georgia coast. (Brachiopods are shelled marine animals that superficially look like clams but are anatomically and functionally quite different.)[11] Predatory gastropods similar to and perhaps ancestral to modern ones were probably responsible for these holes. But these trace fossils are rare compared with modern examples. Over geologic time this scarcity shifted to commonplace, signs that drilling shells and proboscis insertions as a way of life and death escalated as marine predatory snails evolved into more efficient killers. Today moon snails and other naticid gastropods are the most common of such drillers. Other predatory gastropods, such as muricid snails—including the appropriately named Atlantic oyster drill (*Urosalpinx cinerea*)—also scrape holes into shells to get freshly killed meals.[12] Some octopi likewise use drilling as a means of getting past molluscan armor, leaving their own holes in shells, distinctive from those of naticids or muricids.[13]

So let's say you must be reincarnated as a shelled, burrowing molluscan on the Georgia coast. Using what you now know, you might figure the only way to stay safe is to be reborn as a moon snail. Sadly, you would be in for an unpleasant and ironic surprise, as moon snails are also cannibals. Many a time I have picked up a beautifully preserved moon-snail shell on a Georgia beach, turned it to one side, and beheld a hole that could have been drilled only by one of its own kind. Although cannibalism is frowned on in human societies outside of Wall Street, moon snails have no social taboos against eating their own species.[14] Hence if lacking sufficient prey, a moon snail will forcibly invite a brother, sister, cousin, or other relative to dinner. This tactic has its own risks for the attacker, though, as the besieged snail has the same killing implements, meaning the dinner may easily become the diner. A hole in a shell may thus reflect a sibling rivalry turned fatal, but with the instigator left uncertain.

Other animals that find moon snails delectable include crabs and shorebirds, which leave traces of their hankerings.[15] When blue crabs and stone crabs (*Menippe mercenaria*) capture a moon snail with their claws, they hold it with one claw while trying to peel or break its shell near the aperture with the other claw.[16] Telltale signs of a successful crab attack are empty shells with ragged edges. If a moon snail is lucky enough to escape the clutches of a crab, though, its getaway is recorded in its shell as a ragged line from which new shell grew.[17] Alternatively, fights against shorebirds are normally one-sided, as many shorebirds' beaks can easily pierce the stoutest of moon-snail shells. These breaks—evident as oval or circular holes—are much cruder than the carefully carved drill holes of their own species. I have even seen instances of shorebirds using their own form of ichnology, plucking moon snails from the ends of their burrows before cracking them open for a little escargot on the beach.

We can certainly imagine reincarnation for ourselves. But can moon-snail shells actually come back to life? Yes, they can. It is not unusual to see moon-snail shells, with or without drill holes in their sides, sprout legs and go for

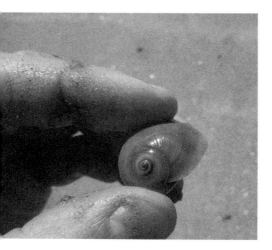

FIGURE 17. Moon snails, beautiful and deadly. *Left*, living moon snail snatched from its shallow burrow on Sapelo Island. *Right*, a very much dead moon snail with a beveled hole that only could have been caused by one of its own, hence a trace of cannibalism.

strolls. When you are on a Georgia coast beach, look for moon-snail shells exceeding posted gastropod speed limits while leaving nicely stitched trackways behind them. This oddity means you just found a hermit crab. Gently pick up the shell, and you may briefly see its resident's four walking legs before they snap back into its adopted home. Hold it a while longer, and the legs may slowly slide out, followed by claws and stalked eyes, the crab seemingly curious at what bequeathed it instant weightlessness. But if an empty moon-snail shell lacks legs, it still may have served as a hermit-crab home in past. The easiest way to discern this past rebirth is to look for a shiny, oval-shaped spot near the shell bottom. This is a trace of the hermit crab, showing where the relatively softer calcite shell dragged along harder quartz sand, wearing it down at that place.[18] Hermit crabs are in a sense, then, carrying out taphonomic experiments for us, showing us what would happen to a moon-snail shell if it did not have the protection afforded by a foot and mucus.

Dead moon-snail shells serve not only as living spaces for hermit crabs but also for many other species that settle on and encrust shells. Among these are colonial animals, such as sponges and bryozoans ("moss animals"), as well as serpulid worms, which make masses of thin, sandy tubes, and

FIGURE 18. Moon-snail shells and their traces, alive and dead. *Left*, shallow burrow of a moon snail, with its maker peeking out of one end. Scale in 0.5 centimeters. *Right*, moon-snail shell that took on a second life after being occupied by a hermit crab, with a wear trace on the bottom of its shell as a telltale sign. Scale in millimeters.

barnacles. These tiny communities often clearly demonstrate an order, showing who colonized first, second, third, and so on. For example, sometimes a drill hole is encrusted by one of these secondary occupiers, literally covering up fratricide by radula. In others one or more of these occupiers cover the worn patch of a hermit crab, which is later exposed and overlapped by a second or third spot made by successive generations of hermit crabs.[19] Each dead moon-snail shell thus informs of lives well beyond that of its original owner, revealing complicated histories in a single shell, analogous to William Blake's beseeching us to consider "a world in a grain of sand." Nonetheless, moon-snail shells bearing new beings can also become battlegrounds again, as predators recognize the new and different lives within. I have seen shells reanimated by hermit crabs that were then targeted by predators, such as ghost crabs. In one instance I found three empty moon-snail shells outside of a ghost crab's burrow entrance, all with worn spots on them, very much looking like a warning to other hermit crabs not to approach.

But enough about moon snails and violence: what about sex? Moon snails do indeed have sex, and the consequences of gastropod lust are evident as so-called sand collars. Mother moon snails make these sandy, stiff (but flexible), semicircular, and ribbonlike structures, which superficially resemble shirt collars.[20] These collars, which may number in the thousands on a Georgia beach at low tide, are egg cases containing the tiny shells of unborn moon-snail larvae, called veligers. Mothers form collars under the sand and offshore, extruding the eggs in a mix of mucus with surrounding sand. The collar turns out curled and potato-chip thin because the jam-like mass flows between the moon snail's shell and the surrounding sand.[21] When the collar stiffens, she pushes it up to the surface, where tides and waves move it elsewhere. As the sandy coating breaks apart, it releases the veligers inside, which start their little lives floating and swimming. Oddly enough, baby food for these future predators is mostly phytoplankton (microscopic algae), meaning the first five to six weeks of their lives are spent as herbivores.[22] The pelagic salad bar stops and the carnage begins when they settle to the ocean bottom, burrow in, and immediately drill into their closest and appropriately sized neighbors. So it goes.

Given the variety and number of traces manifested both by moon snails and the animals that interact with living and dead moon snails, it is not hyperbolic to say these animals contain ichnological multitudes. Despite having seen, held, and studied thousands of moon snails alive and dead, with each newly encountered one I still find myself filled with an appreciation for the myriad connections they make between cycles of life, death, and rebirth: enlightenment latent in a single shell.

10

Rising Seas and Étoufées

I was first introduced to the crayfish of Jekyll Island in 2008 but have yet to actually meet any of them. However surprised I was to learn from other residents on Jekyll that these charismatic freshwater animals lived there, we confirmed their presence in 2008, again in 2012, and on subsequent visits to the island. Still, despite not having seen any of these mysterious dwellers of the underground, I am also sure they are likely living on at least two other Georgia barrier islands, thanks to their distinctive traces and a little bit of detective work.

Crayfish are easy to perceive if you know their traces. For instance, if you see conical towers reminiscent of miniature volcanoes in parts of Georgia and many other places in the southeastern United States, you are probably looking at the handiwork of freshwater crayfish. A few other clues that confirm their crayfish origin are piled balls of sandy mud composing the flanks of these "volcanoes" and nearly perfect circular holes in their centers. Granted, sometimes you may need to lop off their summits to see these holes, but they are there. Then check their ecological context. Are these towers on a floodplain, close to a creek, river, lake, or other freshwater body? If so, congratulations, your hypothesis just went from speculative to firm: these are probably the top portions of crayfish burrows.

The Jekyll Island towers fulfilled all those criteria, and when I saw them in 2008, they were abundantly distributed along the edge of a freshwater wetland in the middle of a maritime forest. They stood out on the flat forest floor as 7–10-centimeter (3–4-inch) tall structures, each about the height of a typical coffee cup. The rounded-oval balls of sediment on their sides were about 1–1.5 centimeters (0.4–0.6 inches) wide, smaller than golf balls but bigger than hazelnuts. They were wider than tall, 10–15 centimeters (4–6 inches) wide at their bases, approaching the width or length of a small paperback book. The first time I saw these towers on Jekyll, the sediment balls were fairly distinct and had not yet collapsed. This overall appearance conveyed how they were made at about the same time and had not appreciably weathered; hence they were still fresh. Knowing a little bit about crayfish behavior, I figured they were constructed just after the latest rainfall on Jekyll, perhaps less than a week before I visited them. All in all, I had little doubt that a thriving population of these crustaceans was living under my feet then.

Although not all crayfish dig burrows, most do, and they adjust the depth of their burrows according to the water table. Similar to many of their crustacean relatives, crayfish—also known as crawfish, crawdads, mud bugs, and other colorful nicknames—have gills, which means they must stay in or near water to stay alive.[1] If the water table drops, they burrow deeper, but if the water table rises, they move their burrows up. For example, where I live in the metropolitan Atlanta area, crayfish towers often pop up in people's backyards the day after a hard rain. (This also means these people should get flood insurance, because their backyards are on a floodplain.) Crayfish dig by using their prominent front claws (chelipeds) as spades, in which they scoop and roll up balls of sediment. They then carry these balls up the burrow shaft and place them outside the top burrow entrance, one chunk at a time. Although burrows are lumpy on the outside, they're smooth on the inside. Burrow interiors are converted into polished tubes by crayfish bodies, which move up and down and back and forth in shafts and tunnels (respectively), resulting in circular cross sections. Crayfish-burrow systems can be complicated, with vertical shafts connecting the surface with underground parts consisting of branching horizontal tunnels and chambers, the latter often hosting many crayfish.[2] Although crayfish-burrow networks are not quite as complex as those of ghost shrimp, they certainly follow a similar schema.

What might seem strange to most people, though, is my not having seen a live crayfish on Jekyll or any other Georgia barrier island. Nonetheless, seeing and documenting their traces is good enough for me to know where they're living and how they're behaving. This again demonstrates one of the many advantages of ichnology: you do not actually have to witness an animal to know it lives there, just so long as it leaves lots of distinctive clues. So upon first seeing these towers and burrow cross-sections on Jekyll Island, I immediately knew they were from crayfish. Previous crayfish researchers and ichnologists affirmed my certainty, having linked these traces to their crustacean makers by describing them in loving detail.

FIGURE 19. Freshwater-crayfish burrows on Georgia barrier islands. *Top*, a muddy, lumpy "volcano" erupts in a wetland surrounded by maritime forest on Jekyll Island, with a perfectly round hole in its center sculpted by crayfish claws. *Bottom*, muddy pellets covering the top of a crayfish burrow, also in a low-lying wetland in a maritime forest on Cumberland Island. These pellets are also the handiwork of a crayfish, which scraped out its burrow and rolled these with its claws before placing them above. Scale in both photos in centimeters.

You also do not have to be a professional scientist to recognize crayfish traces, nor do you necessarily have to travel to far-off lands to see them. In 2012 I saw an example of this connection between traces and tracemakers in my home of Decatur, Georgia. Landscapers working in the city cemetery there had drained its lake, forcing the crayfish (which normally lived on the pond bottom) to burrow down to the water table in a desperate attempt to stay wet. There along the edge of the muddy drained pond were hundreds of towers, many holding crayfish that defiantly waved their claws at anyone walking nearby. The muddy sediments, newly exposed to the air, also preserved hundreds of crayfish trackways as evidence of their out-of-burrow foraging.

I have fond memories of crayfish from my childhood. While growing up in Indiana, where much of my outdoor recreation was devoted to hunting and fishing with my father, I remember seeing them on creek bottoms and being fascinated by their lobsterlike bodies. Little did I know that moving to Georgia in my mid-twenties also meant I had relocated to one of the world's biodiversity hotspots for crayfish. Almost seventy species of crayfish are documented in Georgia alone, with additional species in neighboring Tennessee, Alabama, and the Carolinas.[3]

Freshwater crayfish are geographically scattered elsewhere, with native species in other parts of North America, as well as in South America, Europe, Madagascar, Australia, New Zealand, and New Guinea.[4] This seemingly odd distribution with species separated by oceans is a direct result of plate tectonics, which spread and then isolated ancestral crayfish populations from one another during their evolutionary history. These crustaceans (decapods, more specifically) shared a common ancestor with marine lobsters about 240 million years ago, an age determined by molecular clocks—genetic markers that changed at assumed rates through time—that were integrated with fossil evidence.[5] Other ichnologists and I have also documented trace fossils that look very much like crayfish burrows in Late Triassic rocks from about 200 million years ago, which suggests burrowing began in their lineage early in the Mesozoic era, when dinosaurs were still relatively new innovations on the same landscapes.[6]

In that respect 2008 was a very lucky year for me crayfish-wise, because that was also when I contributed in a small way to better understanding the evolution of burrowing crayfish. In a research article coauthored with six other scientists (three paleontologists and three zoologists), we described fossil burrows in rocks from the Early Cretaceous period (about 115–105 million years ago) of Australia.[7] We also had a few pieces of fossil crayfish previously unknown to science, giving us the privilege of naming a new fossil species, *Palaeoechinastacus australanus* (old spiny crayfish of Australia). As of the time of my writing this sentence, it is still the oldest known fossil crayfish species in the Southern Hemisphere.

In our article we also pointed out how burrowing was an adaptation that likely helped these crayfish survive polar winters in Australia during the Cretaceous period. Australia was then close to the South Pole and still part of a huge southern landmass called Gondwana, which was composed of Australia, Antarctica, Africa, South America, India, and Madagascar.[8] Because burrows protect their dwellers against harsh surface conditions, burrowing in general may have helped crayfish make it past environmental crises in the geologic past.[9] If Southern Hemisphere crayfish evolved first in polar Australia, this hardiness may have enabled them to spread into less hostile freshwater environments of neighboring areas before Gondwana split about a hundred million years ago. Sure enough, New Zealand, South America, and Madagascar have native crayfish related to Australian species.[10] However, if any crayfish made it to Africa or India, they did not leave a fossil record, nor did they survive to modern times.

So now let us move from continental drift and evolution over hundreds of million years to a less grand but still intriguing question: how did freshwater crayfish get onto the Georgia barrier islands? Before answering that I can tell you how they did not get there, which was from people. Because these are burrowing crayfish, and not ones that normally hang out on lake or stream bottoms, humans were less likely to have purposefully introduced them on the islands for aquaculture. Imagine looking at a recipe for crayfish étouffée, seeing that it calls for a pound of crayfish tails, and then having to dig into each individual crayfish burrow to search for a tail or two. Also, excavated burrows may or may not contain a crayfish, because these crafty crustaceans may have dug escape tunnels, allowing them to dash into neighboring burrow systems. This means diggers risk coming up with nothing after hours of labor, and no matter how good the recipe, if this much work is required to make your crayfish dish, you should just make gumbo instead. This dire culinary scenario also explains why crayfish are grown commercially in constructed ponds so they can be more readily harvested, involuntarily donating their bodies to crawfish boils and étouffée.[11]

Another point to remember about crayfish is that they are freshwater-only animals, and most cannot tolerate saltwater immersion, let alone swim long distances through marine waters to reach offshore barrier islands.[12] For freshwater-crayfish species throughout the world, then, the plate-tectonics solution easily explains how they dispersed to places now separated by oceans. But for the much shorter amount of geologic time represented by the Georgia barrier islands, plate tectonics as a process does not apply. Notice I said "islands," which is because I have seen crayfish burrows not only on Jekyll but also on two additional Georgia barrier islands, Cumberland and Sapelo. Moreover, one of the world's experts on crayfish, Horton H. Hobbs Jr., reported crayfish on St. Simons and St. Catherines Islands in the 1940s.[13] Although I don't know if crayfish are still on those islands, their presence in the 1940s suggests they likely lived there well before then. Somehow these freshwater-only animals made their way onto at least four Georgia barrier islands and probably without the help of people.

So what is my explanation for how these crayfish got to the islands? They lived on the islands before they were islands. In other words, present-day crayfish on the islands would have descended from species that originally lived on the mainland part of Georgia, but a few populations were cut off from their original homeland by the last major sea-level rise. This rise started about 11,500 years ago, when the last great ice age of the Pleistocene epoch ended.[14] The global warming that happened then released liquid water from continental glaciers and expanded the seas. This heightening continued until about 5,000 years ago, well before the current rapid (and human-caused) rise in the past few hundred years.

So I would like you to imagine a salty moat filling low areas between what are now the Georgia barrier islands and the rest of Georgia, with crayfish on either side of it, metaphorically waving good-bye to one another with their claws. The crayfish of the Georgia barrier islands would thus represent relics left behind and isolated from ancestral mainland populations. With 5,000–10,000 years having elapsed, they may have even undergone enough genetic drift to become new species or are otherwise well on their way to becoming isolated from their mainland relatives, not just geographically, but also reproductively. But this is admittedly speculation on my part. Like I said, these animals need to be studied thoroughly before anything more can be said about them.

All these crayfish-inspired insights neatly illustrate how our knowledge of the geologic past ties in with the present, as well as how ichnology can be applied to conservation biology. After all, if these barrier-island crayfish are rare species limited only to the Georgia barrier islands, then they need protection to prevent their extinction. If so, then these little muddy crayfish towers exemplify one of the dangers associated with quick and careless development of the islands. What if most people are not aware of the unique plants and animals in these places because at least some of this biodiversity lies below their feet? Without such knowledge unheeded construction may result in destruction by wiping out species that were part of the ecological legacy of the Georgia coast for at least the past 10,000 years. This awareness is one of many reasons why environmental protection of the Georgia barrier islands is still needed, even on developed ones like Jekyll and St. Simons. Fortunately, motivated people are working toward such protection on Jekyll and St. Simons, and most other Georgia barrier islands are under state or federal protection or privately owned as preserves.

What also gives me hope for the crayfish and other wildlife of Jekyll Island is increased ecotourism, highlighted by the success of the Georgia Sea Turtle Center. The center, which opened in 2007, hosts a rehabilitation center for injured turtles, educates the public about sea turtles nesting on the Georgia coast, and helps monitor turtle nests on Jekyll during the nesting season.[15] And just how is this monitoring done? By looking for tracks of the nesting mothers on the beaches of Jekyll during nesting season, demonstrating another example of how ichnology can be used as a tool in conservation and education.

Can a Jekyll Island Crayfish Center be far behind? I wish, but probably not. Still, it is time to start thinking of these animals and other unique species on the Georgia barrier islands (along with their traces, of course) as assets, bragging points that can be used to bolster ecotourism on the coast. The maintenance of barrier-island biodiversity is not just an ethical concern but also an economic one. Biodiversity is a resource that will continue to pay off for as long as species survive and their habitats are protected, while simultaneously feeding our sense of awe at how these species—including burrowing freshwater crayfish—got onto the islands in the first place.

11

Burrowing Wasps and
Baby Dinosaurs

Expect to be surprised. This is my recently adopted can-do attitude whenever I step foot on a developed barrier island of the southeastern U.S. coast
in a quest for animal traces. Once primed by such open-mindedness, I realized that searching beyond the anticipated and listening for the whispers
below the shouts sometimes yield traces of the unexpected. Then I am more
gratified when that previously unsought knowledge reconciles later through
situations far displaced in space and time.

This sort of synchronization happened to me in the summer of 2014, starting with a late May trip to Savannah, Georgia, and its local coastal environments for a few days. Accompanied by my wife, Ruth, the main reason for
our going to Savannah was for a book-related event, but it was really about
beer. With the help of friends there, Ann and Andrew Hartzell, we arranged
for me to give an informal public talk on a previous book I had written about
Georgia-coast ichnology (*Life Traces of the Georgia Coast*) at a local brewpub.[1]
The brewpub had hosted similar events before, but with mine the staff decided to try something different by holding it in their outdoor beer-garden
space. I happily accepted this venue for both its novelty and to decrease my
carbon footprint, upholding a firm moral principle of "think globally, drink
locally." Because we were in Savannah—the setting for the best-selling book
(but not-so-successful movie) *Midnight in the Garden of Good and Evil*—I also
later dubbed the event "Ichnology in the Beer Garden of Good and Evil."[2] The
gathering was mostly a success, in that several dozen people purposefully
attended to hear me speak about my book, whereas others got a side dish of
science with their beer and meals and may have accidentally learned something new. I even sold more than one book afterward, which I happily signed
for people in between sipping a refreshing Hefeweizen.

Given one more night in Savannah and our proximity to the Georgia coast, Ruth, the Hartzells, and I simply had to go to the nearest barrier island the next day, which was Tybee. But one of the main challenges presented by Tybee Island, and one that causes most coastal naturalists to avoid it like an Osmonds reunion tour, is its degree of development. Tybee also attracts large numbers of people, especially on beautiful weekends during the summer. Granted, the development is not so complete that Tybee is devoid of natural environments. But it does have enough paved streets, houses, vacation rentals, hotels, restaurants, shops, and other urban amenities that you can easily forget you are on a barrier island.

Tybee's beaches also differ considerably from those of most Georgia barrier beaches, as parts of it have jetties composed of riprap, consisting of massive boulders of metamorphic rock moved hundreds of miles to the coast. It also has concrete seawalls, fences, and other modifications to its shore. All these unnatural items were placed on Tybee's beaches in a vain attempt to keep sand from moving along the shore as it naturally does.[3] On a barrier island implementing such engineering measures is like telling blood it can circulate only to one part of a body. As a result, beach sands accumulate in some places, erode in other places, and otherwise testify to a losing battle for control. Moreover, at the time of my visit, fences were being used to preserve Tybee's modest coastal dunes. This is a half-buttocked substitute for healthy, well-rooted sea oats and *Spartina* wrack that would do a much better job of holding sand and allowing more to accumulate.[4] Moreover, the sand in those dunes looks odd to anyone acquainted with Georgia coast dunes on undeveloped islands. This is because the sand was dredged from offshore and dumped there for beach "renourishment."[5] Last, and not just to pick on Tybee's beaches, the island lacks anything resembling an old-growth maritime forest or freshwater pond. So, yes, I suppose those cranky and judgmental naturalists have a point.

Ergo, a pessimistic expectation I had before arriving on Tybee is that its only ichnological offerings would be a barrage of human and dog tracks, a tedium punctuated only by human-generated trash, all of which would assault and otherwise insult our senses. Fair or not, this prejudice had kept me away from Tybee while doing field research for *Life Traces of the Georgia Coast*. Likewise, I stayed off St. Simons Island for a while before finally succumbing in 2009, for which I am glad, as I have returned there many times since.

Then again there was the matter of honoring the all-American right to convenience. Tybee Island is only about a twenty-minute drive from downtown Savannah, and we could so easily drive there thanks to a causeway connecting the island to the mainland. Having been to Tybee several times before while leading students on Georgia coast field trips, I also knew plenty could be learned there if I gagged my cynicism. I even had a research question in mind, in which I wondered how many ghost-crab burrows would be in the dunes there compared with undeveloped Georgia barrier islands.

So off we went in the Hartzells' car on a Saturday morning, and within a half hour we were walking on the south end of Tybee, checking out its dunes, beaches, and (of course) traces. Fortuitously enough, my question about the ghost-crab burrows was answered within a few minutes of arriving at the south-end beach. There we spotted a few of their distinctive circular-outlined holes, sand piles outside of the holes, and ghost-crab tracks scribbled on the dunes. Their traces were not nearly as common as on undeveloped islands, but, still, there they were: so far, so good.

Yet this was not the coolest thing we saw, ichnologically speaking. The dunes also had little holes about the width of a pencil, with crescent-shaped openings and fresh sand aprons just outside these holes. I was fairly sure I knew what made these holes, but, as a scientist, I needed more evidence. After I pointed them out to my companions, we stood in one place and waited a few minutes until one of the tracemakers arrived. Hypothesis confirmed. I had predicted these traces were wasp burrows, and after watching several wasps flying around the dunes, landing, walking up to and entering the holes, digging energetically, and emerging, this was all I needed. The wasps were a species of *Stictia* (probably *Stictia carolina*) and were fairly sizable (about 4 centimeters or 1.5 inches long), sporting black-and-yellow abdomens, black thoraxes and heads, and yellow legs. Interestingly (at least to me), species of *Stictia* are sometimes nicknamed "horse-guard wasps" because they prey on horseflies.[6]

To be more specific, though, these wasps were female and making brooding chambers, elliptical hollows at the ends of long (more than 1 meter, or 3.3 feet) inclined tunnels.[7] These chambers serve as little underground nurseries where baby wasps grow up to become big wasps like their mommas and daddies. But for this to happen, the mothers first need to ensure their babies are protected and well fed. So, after making burrows and chambers, wasp mothers hunt down horseflies, sting them into paralysis, carry them back to burrows, and drag their bodies down to the brooding chambers. After

FIGURE 20. A female Carolina sand wasp doing what sand wasps do, which is to burrow and ensure more sand wasps enter the world from below. *Left*, a small crescent-shaped hole in a dune on Tybee Island makes you wonder who made it. *Right*, a sand wasp kicking sand behind it from the front end of the hole is a good clue toward answering that question.

stuffing chambers with horseflies, mother wasps lovingly lay eggs on them, go back up tunnels, seal them, and fly away to dig a new burrow, repeating as needed. Once the eggs hatch, precious little wasp larvae chow down on the still-alive-but-helpless horsefly, grow bigger, and eventually pupate. Once done pupating, they dig their way out of the brooding chamber as adults.[8] Like many insects, such as ground-nesting ants, bees, beetles, cicadas, and more, they were kept safe from egg to adulthood by staying underground.

"Parasitoid" behavior is the term for a predatory life cycle that uses the death of a host as an intermediary for bringing up babies.[9] If this horrific life cycle seems familiar, but you recall learning about it from a fictional setting, that memory was likely implanted by xenomorphs depicted in the classic science-fiction films *Alien* (1979) and *Aliens* (1986). (We will not, however,

mention the sequels or prequels following these two films.) Parasitoid behavior is most common in insects, and especially in wasps, with tens of
thousands of species using some variation on its main theme of having their
children kill their own food.[10] Incidentally, I absolutely adore parasitoid
wasps, and you should too, because not only do they reflect splendid adaptations; they also serve as all-natural and very effective forms of pest control.

With these traces and their tracemakers properly spotted and identified,
I felt much better about our trip to Tybee, boosted as we were by both crab-
and insect-related natural history. Before leaving, we also noted moon-snail
drill holes in shells at the beach, tracks by feral cats (*Felis catus*) in the dunes,
and more drill holes, but in tree trunks. Birds made these traces, specifically
yellow-bellied sapsuckers (*Sphyrapicus varius*). Unlike woodpeckers, their
drilling does not result in immediate insect treats but instead represents
delayed gratification. Freshly dug holes in the trees cause them to "bleed"
sap, which traps insects on the trunk for the sapsuckers to lap up later.[11] So,
yes, it was a good day and place to do ichnology.

Yet this brief visit to Tybee and observing its burrowing wasps potentially
inspired larger questions that tie into evolutionary timescales. For example, how long has parasitoid behavior existed in wasps? Did any of these
earlier wasps also dig burrows with underground brooding chambers for
protecting their young? Did such insects live alongside larger animals, such
as birds, or at least animals that behaved like birds? The short answers to
these questions are more than seventy-five million years ago, yes, and yes.
Furthermore, all these answers were reinforced by real-world examples I experienced only five weeks after that day trip to Tybee and seeing its burrowing wasps in action.

My learning more about life underground way back then through flights
of deep-time fancy began in early July 2014 with a literal flight from Atlanta
to Minneapolis, then another flight to Great Falls, Montana. At the Great
Falls airport, my good friend David Varricchio greeted me there. Dave,
whom I knew as a geology graduate student at the University of Georgia in
the 1980s, is now an accomplished dinosaur paleontologist at Montana State
University in Bozeman. Just before that summer of 2014, he invited me to
work with him—along with a dedicated crew of rubble pickers—at one of the
most famous dinosaur dig sites in the world, "Egg Mountain." Only about an
hour drive west of Great Falls and near the town of Choteau, Egg Mountain
contains many Cretaceous-age fossils preserved in the Two Medicine
Formation, a thick sedimentary-rock sequence composed of sediments laid

down by rivers and lakes about seventy-five million years ago.[12] This formation expresses itself west of Choteau and elsewhere in Montana as rolling, variegated badlands of tectonically tilted beds. For three glorious weeks my job there was to roll out of my sleeping bag; unzip my tent; gaze on beautiful red, white, and purple strata backed by the Rocky Mountains Front Range; and wander those badlands. Each day my main goal was one thing and one thing only: finding trace fossils. It was heavenly.

Why are Egg Mountain and its surrounding area paleontologically famous? Nearly anyone who knows anything about dinosaurs can state confidently that some dinosaurs made nests and took care of their young and that their parenting skills were more like birds and less like most reptiles. If pressed, most dinosaur enthusiasts can further say two dinosaurs—the large ornithopod dinosaur *Maiasaura* and small theropod dinosaur *Troodon*—exemplify this concept. Both dinosaurs lived at the same time and place, seventy-five million years ago in an area near Choteau. This was the same area where a then-young graduate student, John "Jack" Horner, and his friend Bob Makela discovered the first-known dinosaur nests in North America in the late 1970s and early 1980s.[13] Through a series of important fossil discoveries and sleuthing, yielding eggs, hatchlings, partly grown juveniles, and adults of *Maiasaura*, they concluded that *Maiasaura* must have taken care of its young by feeding and raising them in nests for their first few years of life. In short, this was nurturing behavior. Hence this part of Montana is where the concept of dinosaurs as caring parents first developed. Or, one might say, it was the incubator for hypotheses about dinosaurs and their babies.

Only about a fifteen-minute walk from the *Maiasaura* nesting site was another productive dinosaur-nest site, also found by Horner and Makela, the one informally dubbed (and previously mentioned) Egg Mountain. The "Egg" part of the moniker is easy to understand, as it was filled with entire fossil eggs and lots of eggshell fragments.[14] The "Mountain" part, though, is more of an exaggeration, as it is only an isolated, modest hill on the high-plains landscape. This place is famous for its body fossils—dinosaur eggs, eggshells, and bones—but I was also attracted to it for ichnological reasons, summarized by three words: dinosaur nest structure.

Egg Mountain is where the first definite trace fossil of a dinosaur nest was recognized. It was a rimmed depression about the size of a kiddie pool, only shallower, and when Dave discovered this nest in the early 1990s, its rim was composed of hard limestone. But it was originally a soft soil compacted and shaped by either one or both parent dinosaurs. Its identity as a nest structure was cinched by a clutch of about two-dozen eggs in its center, which belonged to the small theropod dinosaur *Troodon*, a parentage later verified by embryonic remains in a few eggs. The width of the nest was also perfect for accommodating an adult *Troodon*, which probably squatted above the egg clutch to protect eggs before they hatched. For me, what made this find even more evocative was knowing how the parent dinosaurs moved the eggs after the mother laid them. These eggs are elongated, which means they would have reclined if laid by a mother *Troodon*. Instead, they were nearly vertical, which means either the mother or father dinosaur manipulated the eggs soon after they emerged from the mother dinosaur and partially buried them upright.[15] This postlaying orientation and the nest structure are both trace fossils pointing toward parental behavior, in which these *Troodon* parents greatly increased the chances their future offspring would hatch.

Now let us leave trace-making dinosaurs for a moment and talk about something that really matters, like trace-making insects. What is well known by those who have worked at Egg Mountain is that the dinosaurs were not alone at their nesting grounds. Just below the dinosaurs' nests, egg clutches, and feet were insects, and plenty of them, as shown by numerous fossil burrows, brooding chambers, and cocoons.[16] The cocoons are exquisitely preserved, with many bearing spiraled silky-weave patterns, and are so common in some strata that you can close your eyes and scoop up handfuls of them. This is why when people mention Egg Mountain and its surrounding paleontological resources, I also think about it as the place where I first learned how insect parenting related to dinosaur parenting.

Other scientists and I made this seemingly unlikely connection by asking a few simple questions. First of all, what animals lived underneath the nests and feet of those dinosaur parents and their babies? Second, what behaviors did these animals express seventy-five million years ago? Third, would the behaviors of these animals resemble those of ones living today, or did they reflect evolutionary dead ends? And, last, did these animals also take care of their young?

It turned out that the most common animals living under the dinosaur nests were not only insects but also wasps. In a 2011 article based on the insect trace fossils in Two Medicine Formation rocks near Egg Mountain, Dave and I concluded that the majority of insect burrows, brooding chambers, and cocoons belonged to burrowing wasps.[17] These insects were making and using the burrows for their reproduction in ways very much like modern wasps, such as horse-guard wasps, do today. They likewise took care of their young by digging out brooding chambers at the end of long, inclined tunnels and presumably took care of their offspring by leaving food in those chambers. Most of the chambers contained cocoons, which further implied that larvae hatched in the chambers, ate whatever food their mother wasps left for them, spun cocoons around themselves once they decided to stop being so larval, and pupated. Once their pupation cycles were finished and they were ready for adulthood, they burst out of cocoons, climbed out of their chambers, and emerged triumphantly on the surface. Once there at least a few were probably eaten by lizards, small mammals, pterosaurs, or baby dinosaurs.

Adding to this gruesome scenario of insect attrition was the fact that most of the fossil cocoons in the Two Medicine Formation show that their occupants did not make it past the pupal stage. This conclusion is easy to make, as some of the outcrops in the Egg Mountain vicinity have thousands of perfectly preserved cocoons, protruding from or falling out of limestone beds as beautiful ellipsoids, yet bearing no sign that an adult insect emerged from them. One of the axioms of paleontology, though, is that each animal's tragedy of the past may someday fulfill a paleontologist's dreams in the future. In this respect, then, these thousands of dead Cretaceous wasps gave me much joy that summer, as I studied these trace fossils for more clues about wasp lives and how they related to ecosystems shared with baby dinosaurs.

In my experience these waspy revelations surprise most people who may not be aware of how many modern insects are descended from lineages that shared the same ecosystems with dinosaurs during the 165-million-year history of the latter animals. Dave's and my research also implies that wherever these insects nested, so did the dinosaurs, which makes good ecological and evolutionary sense. For example, the modern burrowing wasps on Tybee Island and in other places on the Georgia coast dig their brooding chambers well above the high-tide mark, as do ground-nesting shorebirds, such as piping plovers (*Charadrius melodus*), black skimmers (*Rynchops niger*), and American oystercatchers (*Haematopus palliatus*).[18] Very simply, drowned larvae and eggs similarly do not lead to adult wasps and birds, respectively, and any genes encoding such inept behaviors would vanish within just a few generations. Likewise, although the Cretaceous wasps and dinosaurs of the Two Medicine Formation were far from the nearest ocean then and nowhere near a high tide, they still had to ensure their nurseries were above nearby water bodies (lakes and streams) and the local water table.

FIGURE 21. Wasp trace fossils from the Late Cretaceous in the Two Medicine Formation of Montana. *Top*, a fossil burrow with a pupal chamber and cocoon at its end, both filled with sediment that later became rock; scale in centimeters. *Bottom*, fossil insect cocoons from the Two Medicine Formation. The cocoons on the left and right are ichnological two-for-one specials: the left one has a partial burrow attached to it, and the right one has an emergence trace (*top*) from where the adult insect said good-bye to its cocoon seventy-five million years ago.

Did these Cretaceous dinosaurs pick out their nesting places based on where insects nested? Yes, but I doubt they followed burrowing wasps to their nesting grounds, just as I doubt modern shorebirds use wasps and other insects as their cues for picking out their nests. Instead, their co-oc-currence points toward the right ecological conditions for both cohorts of animals in the past and present. So these Cretaceous trace fossils provide a glimpse of the ecology of those places at that time, a window into the past landscapes in which wasps, dinosaurs, and many more animals lived and bred. As we gather more information about the trace fossils of these animals from the high plains of Montana, we cannot help but gain better insights in the cycles of life for Cretaceous insects and the dinosaurs that happened to live in their world. But to truly see how they compare to their descendants— modern wasps and birds—we must also study and learn from their traces too: even those on developed barrier islands, like Tybee.

Part III

Beaks and Bones

12

Erasing the Tracks of a Monster

Life can certainly imitate art, as can life traces. In August 2013, while doing fieldwork on St. Catherines Island, Georgia, I was reminded of such parallels after encountering traces made in the same place by two very different animals: alligators and fiddler crabs. What was unexpected about these traces, though, was how the intersection of their traces also evoked scenes from the then-recent blockbuster summer movie *Pacific Rim* (2013).

For anyone who has not seen *Pacific Rim*, it can be summarized as "massive alien monsters (*kaiju*) emerge from the Pacific Ocean and wreak havoc, but we can't stop them with conventional means, so we make giant robots to fight them."[1] In other words, it was mindless summer-movie fun. Yet a theme expressed early on in the movie—and one that came back to me while doing fieldwork later that summer—was how quickly humanity returned to a sense of normalcy following a lull in *kaiju* attacks, even though the assaults destroyed major coastal cities and killed millions of people. This depiction of collective amnesia reminded me of hurricanes striking coasts and inflicting horrific tolls on human lives and property, such as the Sea Islands Hurricane that hit Georgia in 1893. Yet if no hurricane like it has happened since, then the next thing you know, coastal developers start building condominiums on salt marshes and spits.

But enough about malevolent evil exemplified by *kaiju* and coastal developers. Let us instead turn our minds back to traces. I was on St. Catherines Island for a few days with my wife, Ruth, and an undergraduate student of mine, Meredith Whitten, for field reconnaissance of Meredith's proposed senior honors thesis research. Her chosen research area was along the northeast coast of St. Catherines, and it was covered by storm-washover fans, which are wide, flat, lobe-shaped sandy deposits left by storm waves on barrier islands or other coastal areas.[2] On satellite photos of most barrier islands, these fans show up as white splashes of sand extending past the beach and into more inland places behind coastal dunes. Meredith's research project was to investigate, identify, and map the types of animal traces left on these fans, such as tracks, trails, burrows, scrapings, feces, and more. We figured this information would then tell us more about the distribution and behaviors of animals living in and around these transitional environments between the beach and nearby coastal environments, such as the salt marshes and maritime forests.

Our first morning of field time did not disappoint, as it certainly gave us traces well worth pondering, and one in particular. Within a half hour of walking on a washover fan and studying its surface, we found a trackway left by a big American alligator. Paradoxically, we at first did not see the tracks because they were so large and deep, as we had been focusing our attention on the petite burrows and scrapings of sand fiddler crabs (*Uca pugilator*) and the distinctive, smaller five-toed tracks of raccoons (*Procyon lotor*).

For those of you who have not seen clear examples of alligator footprints, these normally show up as alternating pairs of impressions, with a shorter five-fingered track in front and a longer four-toed one behind.[3] Tracks also often have claw impressions, which connect to the next set of tracks as parallel arcuate lines showing where claws dragged along the ground. These alternating pairs of tracks are separated by a wavy tail-drag trace, which is what I spotted first before noting tracks on either side of it. Also, the rear tracks were longer than my hand, measuring just more than 20 centimeters (8 inches) and thus serving as an indirect warning of the size of their maker. He was probably about 2.5 meters (8 feet) long: not record breaking but large enough to inspire heightened wariness on our part. All these traits and more left no doubt about the identity of their maker. In local parlance, "That ain't no lizard."

I was better acquainted with the alligators of St. Catherines Island than anywhere else because that is where an Emory University colleague of mine, Michael Page, and I had mapped their burrows.[4] Under the right conditions and if necessary, alligators burrow, and they use their burrows as dens for overwintering, raising their young, shielding against fires, and much more.[5] In previous visits to St. Catherines, we saw alligators either in or near their burrows, including some used as nurseries for baby alligators that were also guarded by overprotective mothers. Otherwise these dens were seemingly inert holes in the ground, which we dutifully documented by measuring their dimensions and recording their GPS coordinates.

This particular alligator, however, was not just a data point on a map, as I had seen his distinctive and impressively sized tracks in almost exactly the same place a year before then. Another memorable aspect of his footprints that aided my recollection was their proximity to a salt marsh behind the washover fan. When we expanded our vision to beyond the fan, Ruth, Meredith, and I could see long-established trails cutting through the surrounding salt marshes, all of which were, not coincidentally, the width of a large adult alligator.

Although the conventional wisdom about alligators is that these are freshwater-only animals, their traces continually disagree. Alligators get around and in a wide range of environments.[6] For example, our so-called ecological exception was further verified the next morning when I almost stepped on this same alligator as it laid on a path in the adjacent maritime forest. (I survived.) Over the next couple of days, we saw many more tracks made by alligators of varying sizes going into and out of tidal creeks, salt marshes, and beaches along the ocean-facing side of St. Catherines. Ruth and I have also seen alligator trackways extending for hundreds of meters alongside coastal dunes and coming out of the surf on Sapelo Island and elsewhere (which I'll discuss more soon). This ichnological evidence tells us that alligators occasionally go for long walks along the beach or ocean swims. Far be it for me to discourage midnight skinny-dipping on a deserted Georgia-coast beach, but as a public-service announcement, please be aware that you might have company out there.

We could tell by a few traits of these big tracks, such as their crisp outlines and fine-scale impressions attesting to his reptilian awesomeness, that this specific alligator had probably walked through just after the tide dropped and only a couple of hours before we arrived. But when we looked closer at some of the tracks along the trackway, we were astonished to see that something other than tides had begun erasing them, rendering these big footprints fuzzy and nearly unrecognizable.

The culprits for this eradication were sand fiddler crabs. These little crabs thrive in storm-washover fans and are exceedingly abundant at their edges and close to salt marshes. Additionally, they are industrious burrowers, making J-shaped burrows with circular outlines corresponding to their body widths. These fiddlers make other traces by scraping sandy surfaces outside of their burrows, which they roll up into little sand balls with their claws. The sand contains algae, which the crabs eat; then they deposit the balls as refuse on the surface.[7] In this instance, after a massive alligator

stomped through their neighborhood, they immediately got back to work by digging burrows, scraping the surface, and making sand balls. Within just a few hours of the alligator passing through there, parts of its trackway were obscured. If these portions had been seen in isolation and not connected to clear tracks and the tail-drag trace, I doubt we would have identified the remaining slight depressions as the tracks of a large archosaur.

What was even neater, though, was how some of the fiddler crabs took advantage of homes newly created by this alligator. In at least a few tracks, we could see where smaller crabs occupied holes made by alligator claws. These fiddler-crab homesteads stood out because—like anyone moving into someplace new—they just had to renovate and redecorate. For instance, the edges of some alligator-claw impressions were more rounded than normal, having been sculpted by fiddler claws. Freshly made scrapes and sand balls were also just outside of these modified claw marks, helping to conceal the alligator tracks. The ultimate confirmation of fiddler-crab modification was delivered when I squatted down for a closer look, and a few fiddler crabs looked back at me from their recent refuges. In short, the fiddlers saw these cavities in the sand and, similar to hermit crabs spotting old periwinkle shells weathering out of a relict marsh, said, "Hey, free holes!" So they then moved into these "starter holes," not caring what made them. I have often thought of how small invertebrates sense large vertebrate tracks as suddenly imposed obstacles on their landscapes, but I had never before thought of them taking advantage of these traces for their own well-being.

With such new insights in mind, Meredith continued her field research on the storm-washover fans for the next couple of days, with a focus on raccoon tracks and mapping their raccoon-sized trails moving into and out of the fans. The raccoons, tempted by fiddler-crab feasts each night, were apparently heedless of much larger predators roaming the area looking for mammalian snacks. Indeed, another interesting trace interaction we saw the first

FIGURE 22, OPPOSITE. Fresh tracks of a large American alligator prowling a sandflat on St. Catherines Island. *Top*, right-side tracks accompanied by the alligator's tail-drag trace. But what nefarious nonsense is happening to the tail-drag trace, which is covered by tiny balls of sand? Who made that hole next to the drag mark? And what the heck was a raccoon doing in the neighborhood, leaving its track on the tail drag (*bottom of the photo*)? *Bottom*, later, alligator (tracks). Its right-side footprints are nearly obliterated by sand fiddler crabs burrowing, scraping the surface, and depositing sand balls. Scale in centimeters in both photos.

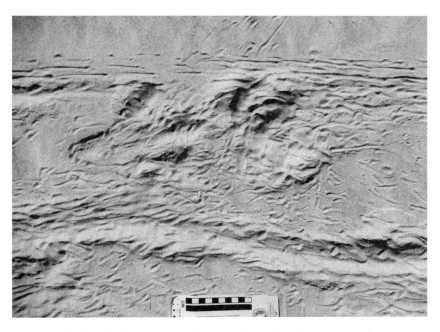

FIGURE 23. Tracks tell of an American alligator on Sapelo Island coming back from a nice, refreshing ocean swim, but ghost crabs then tried to cover its tracks. *Left*, clearly defined (fresh) tracks and tail-drag trace of a large adult alligator moving up a beach and toward the dunes. *Right*, a close look at part of the trackway showing where ghost crabs have blurred the outlines of the alligator tracks with their own. Scale in centimeters. Photo on left by Ruth Schowalter.

day was between those of the raccoon and the alligator. The raccoon's tracks told us it paralleled and crossed the alligator trackway less than an hour after the big saurian passed through, evidently telling itself that the coast was clear.

Almost exactly five years later, in July 2018, Ruth and I had another experience that echoed these lessons of ephemeral moments in the preservational life of an alligator track. Rather than St. Catherines Island, though, this time we were on Sapelo Island for several days, doing field excursions with the express purpose of finding inspiring traces while I was writing this book. We succeeded. On our second morning there, we had just started walking along the north end of Cabretta Beach—immediately south of the relict marsh there—when we were stopped short by not one but two fresh sets of alligator tracks and tail-drag traces coming out of the surf and onto the beach. Even more exciting was their disparately sized footprints, telling us that a different alligator made each trackway.

One of the alligators was medium sized, which based on track-extrapolation experience I estimated was about 1.5 meters (5 feet) long: worthy of respect but not quite intimidating. In contrast, the other alligator was likely larger than the one we had tracked on St. Catherines, approaching 3 meters (10 feet) long. Both had been in the ocean soon after dawn that morning; each was looking for a seafood breakfast but dining separately. The smaller one emerged first, which we discerned by following its tracks up the beach and over the dunes. There behind the dunes, the larger alligator crossed over the smaller one's tracks, perhaps only a half hour later.

Where were they going? Home. More than 50 meters (~165 feet) behind the dunes and in a low area of the back-dune meadow was a freshwater pond. An alligator-worn trail joined the pond to the beach as a path cut through the dunes and plants around the pond. Ruth and I were impressed by how far both alligators traveled, and older trackways in the dunes—pointing to and from the shoreline—hinted at how such seaward sojourns were part of a summertime routine for them. These animals were both swimmers and walkers, performing alligator biathlons every morning while most Georgia coast humans were still in bed.

The most surprising aspect of these trackways, however, was their wide variability in clarity from shore to pond. They ranged from sharply defined footprints—every individual scale outlined, as if etched with a fine-point stylus—to vague undulations that, taken out of context, would not have registered in my consciousness as animal traces, let alone those of large crocodilian predators.

Who or what was responsible for the erasure this time? They weren't sand fiddler crabs but their beach-dwelling cousins: ghost crabs. Where the alligators crossed the high-tide mark and moved into the berm is also where ghost crabs naturally moved down into and across these depressions. For example, put yourself in the place of a ghost crab, and the deep groove offered by an alligator-tail drag with levees on either side of it presents an opportunity to stay low and hidden. Given just a few hours of activity, the repeated movements and collective action of eight-legged ghost crabs across tail-drag traces and footprints resulted in their many curved, scratch-like trackways overprinting those of the giants who preceded them. The small had again reoccupied their rightful place, expunging evidence of the large there before them.

So it's time again to put on my grimy paleontologist hat and ask what we learned from these Georgia coast lessons that might be applied to fossil traces. After all, large four-legged archosaurs analogous to alligators lived through much of the Mesozoic era alongside dinosaurs.[8] We also find their tracks, as well as burrows and other traces of comparatively small invertebrates living in the same places with them.[9] The main lesson I learned is this: any tracks made in the same places as small invertebrates, and especially in areas affected by daily tides, might have been destroyed or otherwise modified immediately by the activities of those much smaller animals. A secondary but related lesson is that large vertebrate tracks can influence the behaviors of smaller invertebrates, causing their traces to interact and blend with one another until they are form a palimpsest of uncertain affinity. A third (but probably not final) lesson is that timing is not quite everything, but it is important for animals making tracks and burrows. For example, had the St. Catherines alligator swam through an inundated washover fan during high tide, it would not have left tracks at all, let alone ones exploitable by the sand fiddlers.

More symbolically, though, these alligator tracks and their erasure by sand fiddler crabs and ghost crabs also conjured thoughts of fictional and real analogues, such as the movie *Pacific Rim* and coastal development, respectively. With regard to the latter, it felt too much like how soon after a hurricane passes through a coastal area as a meteorological "monster," we begin talking about rebuilding in a way that, on the surface, wipes out all evidence that a hurricane ever happened. Yet, unlike fiddler crabs and ghost crabs, we have both memories and records, including the plotted "tracks" of hurricanes. Thanks to science, then, we can predict the onset of future behemoths from offshore. Hence the preceding little ichnological story also feels like a cautionary tale. Pay attention to the tracks while they are still fresh and be wary of those that vanish too quickly.

13

Traces of Toad Toiletry

As an urban dweller who lives most of the year in metropolitan Atlanta, I confess to frequently feeling intense envy of those people who, through sheer number of hours spent in the field on Georgia barrier islands, encounter remarkable traces I have never seen. But I can at least partially forgive them if they act as field partners by proxy by readily reaching out and sharing images of those traces. Of these instances one series of traces in particular was so extraordinary that I suppressed my jealous urges, celebrated the person who found them, and was most grateful for his publicly sharing their specialness.

In the early morning of July 7, 2012, my ichnological benefactor, Gale Bishop, found the intriguing sequence of traces during a morning foray on the dunes and beaches of St. Catherines Island. In Gale's first life he was a geology professor and paleontologist at Georgia Southern University in Statesboro, where he taught for more than thirty years. After that he began his second life, transforming into an indefatigable sea-turtle-nesting monitor on St. Catherines, where he also coordinated a summertime teacher-training program. Among his many duties there, part of his daily routine included looking for mother-turtle traces (trackways and nests) during the nesting season, which in Georgia is from May through September.

During this daily monitoring, and with Gale's eyes well trained for spotting jots and tittles in the sand, he often noticed oddities that likely would have been missed by most people, including me. Among these were traces he photographed on coastal dunes that July morning, which he shared later that same morning on the St. Catherines Island Sea Turtle Program's Facebook page.[1] Gale called me out specifically when he posted these photos on Facebook, and he asked me to tell a story about them. I glanced at the images and gave him my abbreviated take in the comments, which in retrospect read like an elevator pitch for a research article:

Looks like southern toad (*Bufo terrestris*) to me. What's cool is the changes of behavior: hopping, stopping, pooping, and alternate walking (which people don't expect toads to do—but they do).

This was my knee-jerk analysis, which took me a grand total of about a minute to discern and respond. After all, this was Facebook, a forum in which prolonged and deep thought is strongly discouraged. But I also kept in mind that quick, intuitive interpretations later require introspection and self-skepticism, especially when I'm making them. So rather than affirming some inspirational cliché along the lines of "trusting my intuition," I sat down to study the photo with these questions in mind:

- Why did I say "southern toad" (*Anaxyrus* [*Bufo*] *terrestris*) as the tracemaker for the sequence of traces that started from the lower left and extended across the photo?
- What indicates the behaviors I listed and in that order: hopping, stopping, pooping, and alternate walking?
- What signified the animal's changes in behavior, and where did it make these decisions?
- Why did I assume that most people do not expect toads to walk (implying that they think they just hop)?

The first leap in logic—how I knew a southern toad was the tracemaker—was the easiest to make. I had often seen the toads' tracks in sandy patches of maritime forests and coastal dunes. The best-expressed toad tracks show four toes on their front feet and five on the rear, which these possessed.[2] Moreover, the fourth-toe impressions on the rear-feet tracks were elongated, which is typical of toads, frogs, and salamanders. Toads also leave a distinctive bounding pattern, with the front-foot impressions together immediately in front of the rear-foot tracks and the toes of the front feet pointing toward trackway midline. These tracks had all those traits.

Before reaching a final verdict on the tracemaker, though, I tried to be a good scientist and considered other possibilities. After pondering this for a while—perhaps as long as five minutes—the only other similarly sized animal I could imagine making tracks confusable with those of a toad in this environment is a southeastern beach mouse (*Peromyscus polionotus*). But mice mostly gallop when out in the open, with their rear feet moving past their front feet as they move forward.[3] Mouse feet are also very different from toad feet, with toes on both feet all pointing forward, not inward.

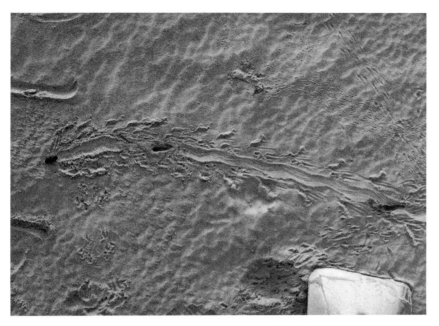

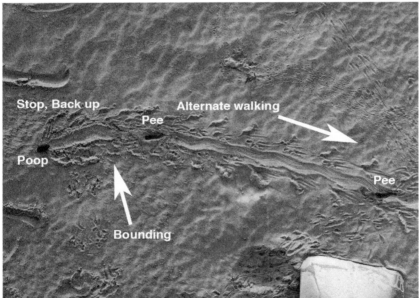

FIGURE 24. A southern toad's toiletry habits written in the sands of a St. Catherines Island beach, with traces unlabeled (*top*) and behaviors labeled (*bottom*); scale is the shovel blade on the lower right. Ghost-crab tracks are also to the upper and lower right. Photos by Gale Bishop, labeling by Anthony J. Martin.

Additionally, mice lack a noticeably longer fourth toe on their rear feet, with the middle toe the longest. So despite dune mice living in the same neighborhood as these tracks, and mice being near the same size, these were not mouse tracks. The only other alternative tracemakers were spadefoot toads (*Scaphiopus holbrookii*) or a same-sized species of frog, such as the southern leopard frog (*Rana sphenocephala*). Yet neither species is as common in coastal dunes as the southern toad. Accordingly, I stuck with my initial identification: southern toad it was.

My second conclusion—the types of behaviors shown and their chronological order—came from first figuring out the direction of travel by the tracemaker, which was evident in the lower left of Gale's photo and toward its middle. This sequence of tracks showed straightforward hopping up to the point where the toad stopped. From there, though, its traces got downright weird. A wide groove extending left of the trackway and past the line of travel must have been made by the posterior-ventral part of the toad's body: colloquially speaking, its butt. This, along with disturbed sand on either side of the groove, show that the toad turned to its right (clockwise) and backed up with shuffling movement. This reversal told of the toad deciding to deposit its scat, a hypothesis supported by the presence of appropriately sized scat. I had seen a similar groove and scat connected to toad tracks before then, and on St. Catherines Island, no less.[4] So this familiar part of the trackway helped me nail down the identity of the tracemaker and gave me good reason to declare, "Hey, I know that turd!"

How about the alternate walking? It turns out that toads do not just hop but also walk: right side, left side, right side, and so on. This pattern—also called "diagonal walking" by trackers—was in the remainder of Gale's photo. When toads do this, the details of their front and rear feet are better defined and you can more clearly see that the front foot registers in front of the rear and closer to the midline of the body. This side-by-side movement is also what imparted a slight sinuosity to the central body drag mark made by the lower (ventral) part of its body, or what one might call its "belly." In my experience most people are surprised to learn that toads can walk like this, which I can attribute only to sample bias. In other words, they have only seen frogs and toads hop away when understandably startled by the approach of large, upright bipeds capable of stepping on them.

But wait! There were also two dark-colored depressions in the center of the alternate-walking trackway. What were these spots? Well, it doesn't take much imagination to figure those out, especially if you've consumed a couple of cups of coffee or beers. These were urination marks, and, even more re-markable, there were two of them in the same trackway. So not only did this toad do Number Two but also Number One—twice. Furthermore, the second mark had less of a stream to it, which makes sense in a way that I hope does not require any more explanation or active demonstrations.

To answer one of the questions previously posed—what signified the changes in behavior?—one has to look for interruptions in the patterns, much like punctuation marks in a sentence. The commas, semicolons, colons, and dashes are all part of a story too, not just the words and their order. In this situation every pause showed this toad was there to make waste, not haste.

Through his series of photos related to this series of traces, Gale also followed a protocol all good trackers do, which is to change his perspective while observing the traces. I likewise teach my students to use this technique when presented with tracks and other traces, that it is a good idea to walk around them. While walking, they see changes in contrast and realize how the direction and angle of light on the traces alters our perceptions of them. At some points while they are circling, a track or other trace may even disappear, then reappear with maximum clarity after just a few more steps.

Now, because I'm also a paleontologist, this interesting series of traces also prompted me to ask this: what if I found this sequence of traces in the fossil record? It would be a fantastic find, worthy of a cover story in *Nature* or at least an appearance on *Ellen*. As of my writing this, I cannot think of any trace fossils like this coming from vertebrates, let alone toads or frogs. So let's go to invertebrate trace fossils for a few examples of a variety of interconnected behaviors representing a short amount of time but preserved in stone.

In 2001 two ichnologists reported a sequence of trace fossils from Pennsylvanian period rocks (more than three hundred million years old) in which a burrowing clam stopped, fed, and burrowed a little more along a definite path, with all of its behaviors clearly represented and joined.[5] The ichnologists who studied this series of trace fossils—Tony Ekdale and Richard Bromley—reckoned these behaviors all happened in less than twenty-four hours; hence the title of their paper was "A Day and a Night in the Life of a Cleft-foot Clam." They also pointed out how any one piece of this clam burrow viewed in isolation would have been given separate names. You see, ichnologists have a sometimes annoying and always confusing practice of giving ichnogenus and ichnospecies names to distinctive trace fossils that do not necessarily reflect their tracemakers. In this instance Ekdale and Bromley said that three names could be applied to the trace fossils made by this one clam, with each a different form made by a different behavior: *Protovirgularia* (burrowing), *Lockeia* (stopping), and *Lophoctenium* (feeding).

Another ichnologist and friend of mine, Andrew "Andy" Rindsberg, and I similarly suggested that an assemblage of trace fossils, represented by many different ichnogenera in Early Silurian rocks (more than four hundred million years old) of Alabama, were all made by the same species of small, short-bodied trilobite.[6] The take-home message of that study, as well as Ekdale's and Bromley's, is that a single species or individual animal can make a large number and variety of traces, especially if it is altering its behavior en route. This also means that ichnological diversity (variety of traces) almost never equals biological diversity (variety of tracemakers).

So let us reconsider the toad traces and think about a "what if" scenario. What if you found this series of traces disconnected from one another: the hopping trackway pattern, the diagonal walking pattern, the urination marks, the groove, and the turd, all found in disparate pieces of rock? Taken separately, such trace fossils likely would be assigned different names. For the sake of silliness, how about *Bufoichnus parallelis, B. alternata, Groovyichnus, Tinklichnus,* and *Poopichnus*? (For the sake of science, please do not use these names beyond an informal and jovial setting.)

Granted, the environment in which Gale noted these traces—coastal dune sands—are not ideal for preserving what he photographed that day. Hence I was very glad he photographed them when he did, because, like many traces left on the surfaces of beaches, berms, and dunes, they likely would have vanished by the end of that same day. Still, other environments, such as pond margins or other more protected areas, might be more forgiving and conducive to fossilization.

If someone does find a fossil analogue to Gale's evocative find, I'll understand their giving a name to each separate part, even if I won't like it. The most important matter, though, is not what you call it, but what it is. And in this case the intriguing story of toiletry habits left in the sand one July morning by a southern toad is worth much more to me than any number of names.

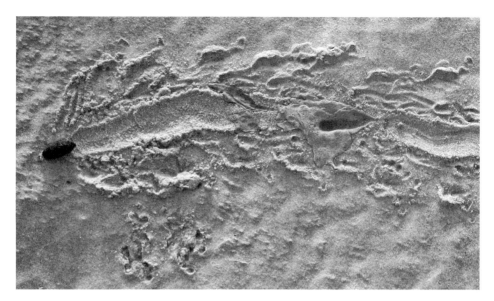

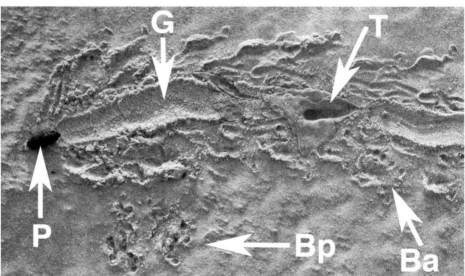

FIGURE 25. Close-up view of the variety of traces made by a southern toad (figure 24) from bounding to stopping, turning, defecating, walking forward, and urinating. Photos by Gale Bishop, labeling by Anthony J. Martin; see text for explanation of abbreviations.

14

Why Do Birds' Tracks Suddenly Appear?

The footprint was familiar, like a face I had seen before but could not quite place. Then I realized to whom it belonged and where I had seen many others like it. It was a bird track, and it was remarkably similar to thousands made daily by the herons, egrets, and other wading birds in the sands and muds of the Georgia coast. The other two tracks near this one were similar in size and shape but not nearly as evocative. This particular footprint conjured an image of a bird slowing its descent from flight, then abruptly halting, planting its foot on a moist, sandy surface.

But this track was from a vastly different time, place, environment, and climate from the modern-day Georgia coast. It was fossilized in an Early Cretaceous (105-million-year-old) sandstone slab collected on the coast of Victoria, Australia, and from an auspiciously named spot, Dinosaur Cove. Moreover, because Australia was close to the South Pole during the Cretaceous period, this track and the other two near it were made in a polar environment.[1] The original environment was not coastal either, but a sandy floodplain in a river valley flooded by meltwaters that flowed and shaped with each spring thaw. Even more incongruously, I first saw this track and its petrified companions on a table in the quietude of a Museum Victoria basement; just outside was the thriving cosmopolitan city of Melbourne, Australia. Mentally and physically, I was about as far away from the Georgia coast as I could be, rendering the track's familiarity both jolting and eerie.

The footprint had four thin toe impressions, looking like the inside of a slightly askew peace sign, with three forwardly pointing and spread widely and one jutting rearward. The forward toes made for a foot length slightly greater than the fingers on my hands. A linear gouge almost as long as the three-toed part of the footprint corresponded with a claw on the rearward-pointing toe, which had also left a faint imprint. In front of the other three digits were piles of sand, only millimeters high but nevertheless there. A thin, sharp claw had neatly bisected another small mound of sand in the center toe impression. This central claw drag was a trace of the bird's next step, in which it pushed against the sand with the bottom of its foot and sliced through the resulting hillock as its foot retracted. The qualities of this fossil track matched those I had seen made by similar-sized birds, such as tricolored herons (*Egretta tricolor*), that landed after flight. I imagined a bird of that size standing on top of the rock slab resting on the table, staring down at me from where I sat.

The long, linear incision behind most of the track was the primary clue to both the identity of its maker and its behavior. This was from its hallux, which in humans is our "big toe," or "digit I" if identified by anatomists. The hallux is the backward-pointing toe on birds that perch, a trait that better allows them to grasp branches in trees.[2] But Cretaceous bird tracks identified elsewhere in the world, such as those from Canada, the United States, Republic of Korea, and China, do not always have a hallux impression.[3] Hence its absence makes separating bird tracks from those of similar-looking nonbird (nonavian) dinosaur tracks much more challenging. This is why hallux impressions are gifts to ichnologists who struggle with distinguishing tracks of the two closely related groups of animals. (In fact, birds *are* dinosaurs, which is why I keep using the cumbersome term "nonavian" when referring to those dinosaurs that were not birds.)[4]

Yet it was not just the hallux impression that convinced me of its avian identity but its lengthiness. This mark was not a mere anatomy lesson but also a window into what that bird was doing one day 105 million years ago: it was flying. The then soft and wet sand behind the main part of the track had been cut by the sharp claw on the bird's hallux, which first contacted the sand before the rest of the foot registered. As this toe slid forward and stopped, the other digits planted, and forward momentum caused their leading edges to push against the sand. This halting mounded the sand in front of these toes just before the bird picked up that foot and took its next step.

The fossil track reminded me of how birds are among my favorite tracemakers of the Georgia barrier islands, a fondness inspired by their great variety there (more than two hundred species), numbers, and diverse behaviors. If pressed to name my absolute favorite types of bird traces, I would not name a specific bird but would immediately say "flying tracks." Granted, "flying tracks" sounds like an oxymoron, as a bird in flight leaves no tracks.

But for those birds where flight is an everyday habit, they must take off and land on something solid, and many of these birds do so on the ground. Ichnologists call such traces "volichnia" (flight traces), which are rare in the fossil record but abundantly represented in soft substrates today wherever flying birds might live.[5] Some of the most redolent of such traces are left in snow, such as those made by owls preying on small mammals. But if you look closely for bird tracks on beaches or river floodplains frequented by flying birds, you will find them there too. Volichnia thus neatly answer the oft-neglected question, why do birds' tracks suddenly appear?

Because these facts of flighted-bird lives are recorded faithfully in the sands and muds of the Georgia barrier islands, I have often delighted in encountering such tracks made by birds varying from sparrows to grackles to gulls to pelicans to great blue herons. Indeed, when I covered this topic in my book *Life Traces of the Georgia Coast* (2013), I had to restrain myself from writing too much about flight traces before moving on to describing other bird traces.[6] Having journeyed down that path before, I will nonetheless describe flight traces again here to ensure the pleasure of recognizing them for yourself.

The first clue a bird was landing or taking off should be discernable from what is *not* there. In short, look for a blank area in mud or sand devoid of tracks just behind a bird trackway, or a trackless place just beyond its last footprints. In both landings and takeoffs, tracks are normally paired side by side. Sometimes, if you're lucky, you'll see both start and end of a landward trek made by one of these feathered visitors. For instance, I once spotted an entire sequence of tracks—from landing to takeoff, made by a common ground dove (*Columbina passerina*) in a back-dune meadow on Sapelo Island—that was probably made in less than a minute. The dove's feet were together when it landed; then it walked in its typical alternating gait, pigeon-toed (or dove-toed, rather) until it encountered a small obstacle, a ghost-crab burrow and its sand apron. There it hesitated briefly and walked around the burrow. Seconds after this change in course, it exited the scene, which doves can do instantly. No more tracks.

Look closer at potential flight tracks, and you will see other details that tell whether a bird was coming down to earth or bidding it good-bye. Landing tracks often have long impressions behind them, or skid marks, that show how the bird decelerated and controlled its fall through a combination of body positioning and calculated flapping.[7] For larger birds with a hallux, such as herons or egrets, these tracks usually leave lengthy scratches from the claws on those toes. While they are landing, one foot plants in

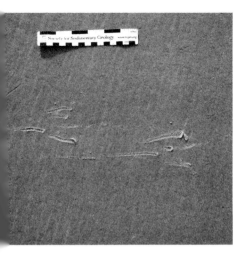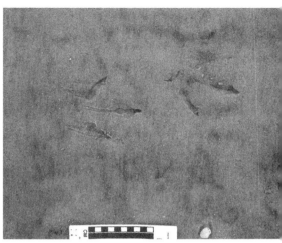

FIGURE 26. Two birds with different feet showing that they stopped flying by landing on a Georgia-coast beach. *Left*, a pair of tracks made by a great egret (*Ardea alba*) made when it came in for a landing on a Jekyll Island beach. *Right*, landing tracks from a laughing gull with nicely defined skid marks on both feet. Scale in centimeters in both photos.

front of the other—either as an offset right-left or left-right pair—and the first track has the longer scratch. One or both footprints also may have mounding of mud or sand in front of its toes prints, caused by the forward momentum of the bird exerting pressure against whatever sediments it encountered. To better visualize what happens with a bird's feet as it lands and takes off, I recommend watching slow-motion videos of flying birds coming in and going away.[8] One in particular that struck me for both its information conveyed and sheer beauty was of a sparrow both landing and taking off. As it approached a surface, it pointed its rear claws toward that surface, which touched first, followed by the forward-pointing toes. Also, one foot barely preceded the other, which in its tracks showed up as a slight offset between the two.

Depending on how fast a bird comes down and taking into account many other factors (like wind direction and speed), this landing pattern could be followed by a hop, or it could just segue into a normal diagonal walking pattern. Also keep in mind that birds with small or absent halluces (plural of hallux), such as sanderlings and plovers, and full webbing between their toes, such as gulls, may just show their forward three digits skidded, leaving no claw traces in the rear part of the tracks.[9]

Takeoff patterns involve opposite movements, in which the feet come together, but the digits dig in and push off. This motion leaves scratches from claws and well-defined mounds of sand or mud behind the digits instead of in front. Such movements can also be ungainly. For instance, I have seen pelican and vulture trackways where they ran or skipped five or six steps, all while flapping their wings before they were aloft, with increasing distances recorded between each successive set of tracks. But sometimes a large bird like a pelican can impress me with its tracks, showing where it successfully accomplished a sudden takeoff from a standing start.

Given these search images, lots of birds, and expansive canvases offered by coastal sands or muds, I am confident that almost everyone should be able to recognize and diagnose their own personally discovered set of "flying tracks." In my case I had not only my little checklist in mind but also years of experience of applying it to Georgia coast bird tracks. When I described bird-flight tracks in my book *Life Traces of the Georgia Coast*, I advised paleontologically inclined readers to apply and test these criteria to fossil bird tracks.[10] But with these tracks from Victoria, I was unexpectedly following my own advice, a situation that fuels uncomfortable feelings in scientists who tend to be overly self-critical of their own work (guilty as charged). Confounding matters, I first saw this fossil track and its two companions in July 2011. My book had not yet been published, nor had I ever written or published any peer-reviewed scientific papers on bird tracks. Sure, I was a living example of what Malcolm Gladwell wrote about in his book *Blink: The Power of Thinking without Thinking* (2005), an expert whose intuition was backed up by a minimum of ten thousand hours of experience (a number that, quite frankly, he must have pulled out of his cloaca).[11] Backing up this intuitive and experience-based conclusion, though, posed a huge challenge, akin to a fledgling trying to decide whether it was time to leave its nest and take a test flight. To continue this metaphor, it is not the ill effects of possible free fall I feared but the academic predators waiting to pounce on an avian-ichnological novice like myself.

Complicating this dilemma was the preservation of the track itself. The bird's right foot made it, but the rock holding these tracks lacked any other evidence of the left foot, let alone the next step taken by the right. All modern volichnia made by landing birds I had seen had paired footprints, right and left together, but slightly offset, with one ahead of the other. But the slab of rock had no track behind this right-foot impression, and it was broken along the front edge of the foot's middle digit. If this bird had landed with the right foot first—which, based on the length of the hallux imprint, I think it did—then the left foot would have been more than a track length ahead of the right. If so, it may be gone forever, taken away by the same coastal erosion of the Victoria coast that delivered the surviving tracks to their discoverers, who arrived just in time to save them.

Who discovered these tracks? Not me. Instead, the invaluable, indispensable, and intrepid allies of desk-bound, exam-grading, lab-teaching, and meeting-imprisoned academic paleontologists everywhere found them: volunteers. On November 29, 2010, Museum Victoria volunteers Sean Wright and Alan Tait were walking along Dinosaur Cove, scouting for dinosaur bones along its rugged, rocky shore. This place was so dubbed not because its outline resembled a *Stegosaurus* or other dinosaur-themed pareidolia but because it was the same place where most of the dinosaur bones in southern Australia were originally recovered. Excavated during the 1980s–1990s, Dinosaur Cove—which is about a three-hour drive west of Melbourne—was among the most logistically difficult dinosaur dig sites in the world, as described by paleontologists Tom Rich and Patricia "Pat" Vickers-Rich in their 2000 book, *Dinosaurs of Darkness*.[12] It and another site about a two-hour drive east of Melbourne, nicknamed Dinosaur Dreaming, have produced the most complete assemblage of polar-dinosaur bones in the Southern Hemisphere.

This bit of paleontological legacy meant that Wright and Tait were not searching randomly along the coast but instead were looking for rocks containing fossil bones that had eroded from the coastal outcrop. Instead of bones, though, Wright spotted a flat sandstone slab bearing three-toed patterns of fossil tracks among the boulders and cobbles in the energetic surf zone there. With this discovery Dinosaur Cove was suddenly and inadvertently added to a very short list of Cretaceous track sites in southern Australia. At the time Wright and Tait figured these trace fossils were probably footprints of nonavian dinosaurs, such as theropods or ornithopods, both of which make three-toed tracks. When Tom Rich emailed me photos soon after Wright and Tait's discovery, I confirmed that they were indeed tracks. I also said they looked a lot like the theropod-dinosaur tracks I had discovered in 2010, in rocks of the same age about ten kilometers (six miles) east of Dinosaur Cove.[13]

On March 31, 2011, a little more than four months after Wright and Tait's discovery of the tracks, Tait went back to Dinosaur Cove by himself with hand tools and a backpack. He then broke the slab into four large pieces so they could be transported on foot, which he did, with all forty-five kilograms (a hundred pounds) of rock on his back. For anyone who has hiked into and out of Dinosaur Cove (and I have several times), this was a remarkable one-person recovery effort, one that some people might justifiably term as, well, loony.

Fortunately, Tait's daftness paid off big time. The bird tracks had come in for a second landing on the rocky shore of Dinosaur Cove, having fallen off the outcrop as a consequence of coastal erosion. Fortunately, Tom Rich recognized their origin. The rock was originally part of a sandstone bed just above the Slippery Rocks Tunnel site, where he, Pat, and many volunteers dug, broke, blasted, sifted, cursed, and otherwise labored in their quest to collect the dinosaur remains there in the 1980s and 1990s. Fittingly, we could say with great confidence that living animals were walking in the same place in this polar river valley where the bones of their possible ancestors were buried. Then, once I looked more closely at the tracks later in 2011, I was happy to confirm that one of those living animals was a flying bird.

With such a great discovery, it was obviously time for us to contact the press, post the news on social media, and otherwise breathlessly report that we had the oldest bird track in Australia. Oh, that's right. In all the excitement about describing tracks of flying birds from the Georgia coast, I almost forgot to mention that not only was this the track of a relatively large flying Cretaceous bird but also that it was the oldest known from the continent of Australia and one of the oldest in the Southern Hemisphere. This was a significant discovery, one worthy of shouting out to the world.

Except that, no, publicizing it would have been totally wrong and would have served as a great example of how science is not done. This discovery first had to go through peer review, which meant that no matter how confident I was about the identity of the tracks, my description and interpretation of them in a peer-reviewed publication and later acceptance by the rest of the paleontological community was not guaranteed. Also recall that despite all of my experience with modern bird tracks, I had never published an article about fossil bird tracks. So to add more credibility to whatever article came out of this find, I asked Pat Vickers-Rich and Tom Rich to coauthor it with me and was delighted when they accepted. Sedimentologist Mike Hall of Monash University also later joined us as a coauthor, adding considerable Australian expertise to the article.

To make an already-long story much briefer, a year and a half went by before the paper was finally accepted and published online in 2013 in the journal *Palaeontology*.[14] Peer review on this paper was tough, rivaling the most

FIGURE 27. The recorded moments of birds coming in for a landing but separated by more than a hundred million years. *Left*, right footprint of great egret, with a long scratch made by its rear toe (hallux) when the bird stopped abruptly. *Right*, right footprint track of an Early Cretaceous bird with a similarly long scratch made by its hallux, preserved in the top of a sandstone bed from Victoria, Australia.

challenging of my academic career. Different versions of the paper went through two rounds of review with four different reviewers and two different editors. Two of the reviewers were anonymous, which meant I could not contact them directly with any questions, although two of them graciously revealed their identities (thank you, Jenni Scott and Matteo Belvedere). Having become so discouraged by negative and demeaning comments from the anonymous reviewers, I almost gave up on the article several times. But in such dark moments I reminded myself of an important affirmation: all the reviewers agreed we had fossil bird tracks, that they were made by relatively large birds (which were rare a hundred million years ago), and that these were the oldest known in Australia. That kept me going.

The attentive reader may have noticed I just said "tracks," as in plural. A great benefit of the sometimes-humiliating scrutiny of the reviewers was that several pointed to the track just left of the "landing" track and said, "There's another one." Although I originally thought this track was from a dinosaur that was not a bird, I reconsidered it and realized they were correct. This track was from a bird's left foot, one with a foot close in size and form to the other one. But this track had a much less obvious and "not-flying, just walking" hallux impression, so understated in its expression that I missed it. One of the more interesting traits of this track, though, was how one of its digits flexed as the foot moved against the sand, leaving a curved impression.

The third track, however, presented a dilemma, as it possessed qualities of a thin-toed nonavian theropod. Think of something like an oviraptorid or ornithomimid, feathered dinosaurs that at first glance would have looked like birds to you and me, but were not. Yet this footprint also could have been that of a bird, in which its hallux just did not register on the sand in that one step. I also considered that it represented a double print, where the foot registered twice from planting, lifting, and planting again in the same place, distorting its features. So just to be conservative and err on the side of caution, we concluded it was probably from a nonavian theropod dinosaur, but we remained open to the possibility that it was made by a bird too.

Could we all be wrong, and none of these tracks are from birds but instead are from some theropod dinosaurs very close to birds in their foot anatomy? That's conceivable, but not likely at this point. Could I be wrong about taking just one track on a small, broken sandstone slab and interpreting it as evidence of flight? Again, that is possible. Alternate explanations include that the bird just hopped—perhaps with a flap or two—before landing, or its foot slipped on wet sand as it was walking forward, recording a fossil "oops" moment. But in my experience with modern birds, pratfall traces are even more rare than flying traces. Could Cretaceous birds in polar Australia have been clumsier than those today, hence slipping tracks would have been more common? Now, that's just silly.

So let us allow ourselves to celebrate the many reasons why this was a significant find. These were, in no particular order, the oldest-known bird tracks in Australia; the only Early Cretaceous bird tracks in the Southern Hemisphere; evidence of egret- or heron-sized birds in a polar environment during the Cretaceous; evidence for flight in an Early Cretaceous bird track, one of the few examples known in the world; the first vertebrate tracks, accompanied by the first-known nonavian dinosaur track, from Dinosaur Cove, a place previously famed for its dinosaur bones.[15] These tracks also provided more evidence for Early Cretaceous birds in Australia, supplementing rare body parts found thus far, including just a few feathers and a single wishbone from Victoria. If anyone had used this bone to make a wish for more evidence of Cretaceous birds in Australia, then it came true.

All in all, it might be just two fossil tracks, but those two tracks made the fossil record for the birds on an entire continent and the rest of the Southern Hemisphere a little better. Still, I also maintain that I would not have recognized them if not for the training bequeathed by thousands of birds leaving their tracks on the Georgia coast. Now that these fossil flying tracks of the past have landed into our present consciousness, it is only appropriate for us to allow our imaginations to take off and find more.

15

Traces of the Red Queen

The herring gull looked peaceful on that beach, lying on its left side with its eyes closed. Yet it was a permanent quiescence, as only its head was there. Disembodied as it was, it otherwise stuck out from its surroundings as a white spot with a red edge perched on top of a pile of dull-brown and similarly lifeless cordgrass. The torso so recently connected to it was nowhere to be seen, and I could find no tracks belonging to either the gull or any other animal around it. It looked almost as if it had been placed there as a macabre object of art, ready for erudite admirers—wineglasses in hand, pinky fingers extended—to comment on its broader themes and nuanced metaphors. To an ichnologist, though, it also spoke of a sudden death, one likely dealt by a ruthless predator.

The place where I saw this gruesome sign was on Wassaw Island, Georgia. Wassaw is rather small compared with its companion islands of Ossabaw to its south and Tybee to its north, but qualitatively it is one of the most interesting on the Georgia coast. For one, it is the only island on the coast that was never logged or otherwise developed by Europeans or Americans.[1] As a result, it retains a more primitive feel compared with most other Georgia islands; hence, while treading on its shores or interior, I always feel as I have stepped back about a thousand years. Adding to Wassaw's sense of remoteness from the modern world, you must take a boat to get there, which sometimes requires winding through a complicated network of salt-marsh tidal creeks between the mainland and the island. To my inner child, going to Wassaw never fails to feel like an adventure.

On this particular visit I had a boatload of university students with me to better cultivate that youthful sense of wonder, all of us led by our captain and guide, John "Crawfish" Crawford. John is one of the most experienced naturalists on the Georgia coast, and for this trip, in February 2014,

FIGURE 28. Head of a herring gull missing the rest of its body while lying on *Spartina* rack on Wassaw Island.

he was working as a marine educator for the University of Georgia Marine Extension on Skidaway Island. Our field-trip group was going to Wassaw so we would learn about the unique natural history there; then the next day I would contrast this experience for the students by taking them to far-more-developed Tybee Island.

Within minutes of arriving on the sandy eastern shore of Wassaw, this beheaded gull presented a little mystery for us. Because of the intact environments and general lack of human influence on Wassaw, though, I was not surprised to see something there worthy of a Washington Irving story. As mentioned before, footprints and the rest of its body were not visible, nor were any droplets of blood around its head. Moreover, its dry feathers and the freshness of its fatal wound—a clean severing of its neck vertebrate—also meant it had not washed up on shore after floating at sea. Where did it die, and how did it get there? After ruling out the land and sea, I looked above the beach and realized that the attack must have been delivered up there, in the air. We then imagined what could have possessed the bulk, ferocity, and other means to chop through a herring gull's neck in flight. The list of suspects was a short one, and we quickly narrowed it down to one: a bald eagle (*Haliaeetus leucocephalus*).

Our hypothesis was not far-fetched, as bald eagles do not just eat fish but also kill and eat other waterfowl, including gulls.[2] This meant the gull head we saw that morning was very likely a result of bird-on-bird predation, in which an eagle attacked and dispatched a gull in midair. In this scenario the rest of the body was taken elsewhere and eaten, but the head may have fallen straight down from above. If we extended the evidence for this attack a bit further into the evolutionary pasts of these birds, however, it also reflects on times when their nonavian dinosaur ancestors killed and were killed by similar behaviors.

Most such lethal confrontations of the geologic past happened on or near the ground. How did birds evolve flight from nonflighted theropod ancestors, allowing some of them to eventually prey on those from the same lineage? No doubt, one of many selection pressures exerted on nonavian dinosaurs during their evolutionary history was predation. Any means for increasing the chances of escape from predators also bestowed a greater probability for passing on genes favoring that escaping trait to the next generation of not-quite-flighted dinosaurs.[3] Of course, flight evolved for many uses in birds today, and making quick getaways from mortal peril still works well for most species. Yet flight has also been used as a means for enhancing predation in birds that slay other birds, exerting new and different selection pressures on their prey.

This example of an evolutionary back-and-forth "arms race" between predators and prey is sometimes called the Red Queen hypothesis.[4] First proposed by evolutionary biologist Leigh Van Valen, the Red Queen hypothesis is named after Lewis Carroll's character in his book *Through the Looking-Glass, and What Alice Found There*, the sequel to his better-known title, *Alice's Adventures in Wonderland*.[5] Only now, I will change the Red Queen's line (said to a fleeing Alice) about running in place:

> Now, here, you see, it takes all the running you can do to keep in the same place.

to a more avian-appropriate version:

> Now, here, you see, it takes all the flying you can do to keep in the same place.

Nevertheless, in this Georgia coast example, a more appropriate literary allusion would have been that of the Queen of Hearts from *Alice's Adventures in Wonderland*, a decapitating character famous for uttering the line, "Off with their heads!" In this sense the Red Queen and Queen of Hearts met in this example of the arms race between flying predators and prey.

Will this "Red Queen of Hearts" scenario happen again during eagle and gull conflicts? Yes—that is, unless the gulls' descendants adapt, meaning the eagles' descendants may likewise follow and adapt accordingly. And on it goes, this evolution of the now blending with the then, a reminder that days of the dead affect those of the living, as well as those not yet alive.

16

Marine Moles and Mistaken Science

One of the few expectations on the first day of fieldwork for any natural scientist, no matter what type of science is being attempted, is that the best-laid plans will go awry. Sometimes these mishaps are unpleasant ones, such as finding out too late that the fuel gauge in a field vehicle you've driven to a remote place is broken. Other times you make a fantastic discovery, such as a new species of spider, a previously undocumented invasive plant, or a fossil footprint, but then you realize you left your camera back at your campsite, or you brought your camera but forgot to charge it or put in a memory card. But sometimes the mistake is more delayed, begun by observing something new to you in the field, something that makes you scratch your head and say, "What the heck is that?" or more profane variations on that sentiment. You then make a wrong assumption about it, and years pass before you realize (with much embarrassment) just how wrong you were.

The last of those three scenarios happened to me on Sapelo Island, Georgia, starting in June 2004 and ending in embarrassment several years later. My wife, Ruth, was with me during that visit, and we had just arrived on the island the previous afternoon, having stayed overnight at the University of Georgia (Athens) Marine Institute. We decided our first full morning in the field would be at Nannygoat Beach on the south end of Sapelo, which is a five-minute drive or twenty-minute walk from the institute. We drove a field vehicle there (the gas gauge and everything else worked), parked, and took a boardwalk over the coastal dunes to the beach. Our elevated view from the boardwalk afforded a good look at many insect, ghost-crab, bird, and mammal tracks made in the early morning. Circular holes dug by ghost crabs punctured the dunes, and sand aprons composed of still-moist sand were next to these burrow entrances. Crisply defined ghost-crab tracks danced across the rippled dune surfaces, although early morning sea breezes had just begun to blur these.

At some point after walking onto the beach, we saw traces I had not noticed in previous visits to Sapelo, ones I have seen infrequently there and on other Georgia barrier islands since. These oddities were sandy ridges about fifteen to twenty centimeters (six to eight inches) wide, many meters long, and slightly sinuous to meandering. They extended from the dunes, into and through the berm, and down to the high-tide mark, where they ended abruptly. Although a few ridges crossed one another, they rarely branched, and, if they did, the branches were quite short, only about ten to fifteen centimeters (four to six inches). In places where separate ridges intersected, they lacked enlargements (roundabouts) that normally would accommodate changes in direction by their maker.

Curious, we followed these ridges back to the dunes, where they seemingly originated from unseen places below the sandy surfaces. We investigated further by carefully cutting through a few to see what they looked like inside. Once dissected, they were revealed as tunnels with circular cross-sections about five centimeters (two inches) wide, or slightly wider than a U.S. dollar coin. They were mostly hollow, and only occasionally did we encounter a plug of sand interrupting tunnel interiors, a supposition supported by ridges that had collapsed into underlying voids. A few also had rounded ends or terminated into larger, raised, elliptically shaped hills.

Ruth and I agreed these tunnels were burrows, instead of some random features made by the winds, tides, or waves. But burrows of what? Clearly their makers were impressive diggers, capable of plowing through meters of sand. Their bodies also were probably just a little narrower than the burrow interiors, which helped us to think about body sizes. Then we considered where we were—dunes and beach—and what animals were the most likely ones to burrow in these environments.

A process of elimination, guided first by determining what they were not, was a good way to start figuring out their potential makers. For example, there was no way these burrows were from insects, such as beetle larvae, ant-lion larvae, or mole crickets (*Scapteriscus* spp.), because they were just too big. Most insects also have a tough time handling salinity.[1] So if they burrowed to the surf zone and its saturated salty sand, they should have been aversively impelled to abruptly turn around instead of moving ahead. Small mammals—like southeastern beach mice—did not seem like good

FIGURE 29. Long, shallow, meandering tunnels in very different environments on Sapelo Island, but made by the same species, the eastern mole. *Left*, tunnel connecting coastal dunes and the high-tide mark on a beach. *Right*, tunnel describing a hairpin turn on a sandy road in a maritime forest on the north end of Sapelo Island, far from today's ocean.

133

FIGURE 30. Ghost crabs trying too hard to be misidentified as the makers of these mole tunnels by gullible ichnologists (like me), adding their own burrows and tracks to the mole burrows. Scale in centimeters.

candidates for the burrow makers either. Beach-mouse burrows are totally different from what we were seeing, and their burrows tend to stay in the dunes and back-dune meadows, not travel all the way down to the intertidal zone.[2] Also, like insects, mice do not like marine-seasoned water. Even if they temporarily tolerated it, they certainly would not have continued to burrow through moist sand for such distances.

This negation of possible burrowers led to my initial hypothesis, which was that these burrows were made by one of the most common larger burrowing animals in the area. Second, it also had to be an animal comfortable in dune, berm, and beach environments with wet sand. These could only be from ghost crabs, I thought, an explanation supported by undoubted ghost-crab burrows that perfectly intersected these tunnels, which in turn were accompanied by undoubted ghost-crab tracks. The circumstantial evidence and coincidence of it all was just too good to ignore.

End of story, right? Well, no. A lot of other scientists and I have said this before, but it bears repeating: part of how science works is that in its practice we do not prove, we disprove. I knew the "ghost crabs burrowing horizontally through meters of sand from the dunes to the beach" hypothesis was a shaky one. It bothered me that it just did not ring true, and after a while it began to inwardly mock. This self-directed ridicule motivated me to read as much as possible about ghost-crab burrowing behaviors. I thought I already knew plenty about this subject but nonetheless was willing to acknowledge that there might be some holes in my learning (get it—holes?) that needed filling (get it—filling?). So I dug a little deeper in my groundbreaking research. (Sorry.)

Given so much foreshadowing, nearly everyone should have surmised what happened next. Not a single peer-reviewed reference mentioned ghost crabs digging meters-long shallow and meandering tunnels from the dunes to the beach. Either I was wrong, or I had documented a previously unknown and spectacularly unusual behavior in this well-studied species, akin to discovering that alligators climb trees. A single slice by Occam's razor simply said, "I'm wrong."[3] If not a ghost crab, then, what else could make meters-long horizontal burrows of the diameter we had seen? This is when I began to reconsider my original rejection of small mammals as possible tracemakers.

Probably the largest contributing factor to my mistaken interpretation, and the most interesting, is that Ruth and I had been primed because of where we were. In other words, our initial mystification about these traces stemmed from their environmental context. Had we seen these burrows winding down a sandy road in the middle of a maritime forest on Sapelo Island, we would not have hesitated to say the word "mole." Yet because we saw exactly the same types of burrows in coastal dunes and beaches, we said, "It must be something else."

Do moles, like other mammals, go to the beach? A trip back to the published literature said yes, further supporting the mole hypothesis while also serving up a big slice of humble pie. I was discomfited to find out these same ridges, tunnels, and mounds had been described and interpreted as mole burrows in an article published in 1986.[4] Even more mortifying? My dissertation adviser, Robert "Bob" Frey, was the first author on the article. It was published while I was doing my dissertation work with him at the University of Georgia, and I had read the article years ago but did not remember the part about mole traces. It was as if these burrows were telling me, "Why don't you just go back to school and learn something, young man?"

Okay, so these were mole burrows. Case closed. Now that we've identified them, we can stop thinking about them and go on to study something else.

But that is not science either, is it? This one answer—mole burrows—actually inspires many other questions about them, which could lead to heaps more science.

The first question that probably comes to mind is, Which moles made these burrows? The Georgia barrier islands have two documented species of moles, the eastern mole (*Scalopus aquaticus*) and star-nosed mole (*Condylura cristata*).[5] Of these two, eastern moles are relatively common on island interiors, whereas star-nosed moles are either rare or locally extinct. But star-nosed moles are also more comfortable next to water bodies than eastern moles, and they often seek out underwater prey.[6] Could these traces actually signal the presence of rare star-nosed moles in dune and beach environments? In their 1986 article Frey and his coauthor, George Pemberton, originally interpreted these burrows as those of eastern moles, but they did not eliminate the possibility of star-nosed moles as the tracemakers either. I personally would be thrilled if someone, someday, linked this species to these traces. Yet the easier explanation for now and the one I conditionally accept is that these burrows are from the more common eastern moles.

Another question—and more of a big-picture one—concerns the evolutionary history of moles on the Georgia barrier islands. Are these moles like the crayfish on some islands, in that they descended from populations isolated from the mainland as much as ten thousand years ago by the post-Pleistocene sea-level rise? Or do they reflect a more recent stock that somehow made its way to the islands? A genetic study of moles on the Georgia barrier islands compared with those on the mainland would probably partially answer these questions. But given the state of scientific funding for such curiosity-driven research, I have little confidence in that happening any time soon.

Yet another question I had was about their diet. What were these moles eating? Just like how artists, musicians, and writers do not create simply for exposure, moles do not burrow for exercise. While digging, they are also voraciously chowing down on any invertebrate encountered in the subsurface. But what would they eat in beach sands? As many shorebirds would remind you if they could, Georgia's sandy beaches are full of tasty amphipods, which would likely substitute for a mole's typical menu of earthworms and insects in terrestrial environments. Yet as far as I can find in the scientific literature, no one has documented mole stomach contents or scat from coastal environments to test whether or not these small crustaceans are their main prey in these places.[7] Perhaps the mole geneticists and mole dieticians need to join forces in such studies, united through their love of subterranean small mammals. And perhaps their research would even provide incentive for remaking classic films such as *The Mole People* (1956), only this time with scientists as the true heroes.[8]

Scientific and cinematic aspirations aside, one of the little mysteries for me about these burrows concerned the ones that ended at the surf zone. What happened to the moles once their burrows stopped? Did they turn around and burrow back, or did they go for a swim in the open ocean? The latter scenario is actually not so far-fetched, as moles (and especially star-nosed moles) are excellent swimmers.[9] But how often would they purpose-fully go into the water? If so, would we find mole tracks on beaches coming out of the surf and connecting to their burrows? As much as I would love to find cute little mole tracks on a beach, though, I have a hunch these moles were not swimming at all but instead burrowed during low tide. In this sce-nario the tunnels that reached the farthest down the beach were abandoned and eroded by the next rising tide, meaning their makers would not neces-sarily need to have swum.

How common (or rare) are these burrows in Georgia beaches? Another potential pitfall of scientific thinking is sample bias. In other words, just because I originally perceived these burrows as rare does not mean they ac-tually are uncommon. Sure, I wrote a doorstopper of a book about Georgia coast traces and tracemakers, I've done fieldwork on nearly all the islands since 1998, and I repeatedly taught a college-level class about barrier islands with a special focus on the Georgia coast. So I can claim some experience and expertise, and my opining about the relative rarity of mole burrows on the Georgia islands does not make me the natural-history equivalent of a crazy uncle sharing unsolicited views at Thanksgiving dinner. Yet I do not live on the Georgia barrier islands, nor have I spent more than a week con-tinuously on any of them. The people who could really answer this question are the keenly observant naturalists who live on the islands or otherwise spend much time there. Also, because I suspected these burrows are actually much more common than originally supposed, and I know what to look for, I have seen more ridges, tunnels, and mounds along Georgia beaches, which I made sure to photograph or otherwise document each time.

A last question, and one every geologist might ask, is this: would such burrows preserve in the geologic record and be recognizable as burrows millions of years later? They probably would, especially if they were made in dunes and their formerly open tunnels were filled with differently colored or textured sand. But I also bet that nearly every paleontologist or geologist would make the same mistake I did and simply assume a burrowing crab or some other invertebrate animal was the tracemaker, not a small mammal. Geologists would be further led down a foolish path if these fossilized mole tunnels intersected with genuine fossilized ghost-crab burrows. Such Frankenstein-monster reconstructions would constitute excellent examples of what ichnologists call "composite traces," ones made by more than one species of animal.[10] Why did modern crabs burrow into the mole tunnels? They did so because it was easier. After all, the moles left hollow spaces and loosened sand over wide areas, which invite ghost crabs to exploit these disturbed areas. Regardless, I doubt many geologists would think of a small terrestrial mammal as a tracemaker for such lengthy burrows in sedimentary rocks in coastal dunes, berms, and beaches. Still, I would love for my colleagues to prove me wrong, and I hope my writing about my mistakes and lessons learned here will help prevent future confusion for them.

In short, my example of making a crab burrow out of a mole tunnel serves as a cautionary tale of how where we are when making observations in the field can influence our perceptions. But it also goes to show us how our admiration for what we observe in natural environments can be renewed and encouraged by daring to be wrong and then learning from our mistakes.

17

Tracking That Is Otterly Delightful

Writing about a place, its environments, and the plants and animals of those environments is challenging enough in itself. Yet to write about that place and what lives there, but without actually being there, feels almost fraudulent. Given a specific place, I could certainly read everything ever published about it, watch documentaries or other videos about it, carefully study 3-D virtual-reality images of its landscapes, interview people who have spent much time there, and (most radically of all) read books. This is how most of us learn, particularly in our Internet-dominated world: we gather information vicariously without experiencing it directly. But then does my writing about those topics compare to Plato's allegory, in which I am reduced to describing shadows cast on the walls of a cave as representing what is actually there? Could I not gain better insights by sliding down its stalagmites or trying to climb up its stalactites?

For instance, if you show someone a photograph of a Georgia coast beach and ask what is there when they have never visited it, it would be all too easy to say, with conviction and finality, "Looks like a beach." But when standing on that beach, I am much more likely to say that I see fine quartz and heavy-mineral sand that originally composed much larger rocks in the Appalachian Mountains. I see that same sand blowing down a long beach but pausing to form wind ripples or piling onto racks of dead cordgrass washed up by a high tide. I see a river otter (*Lutra canadensis*) galloping alongside the surf, slowing to a lope, then a trot, then back to a lope and a gallop. I see it enjoying its time on that beach while moving around obstacles large and small, including racks of cordgrass and their protodunes. I see a brief rain shower less than two hours after the otter has left the beach, and the wind gusting afterward. All in all, there is something different about having been in that place that causes me to be a little more observant and philosophical.

Then add traces to the mix, and the stories that emerge become more alive and real.

Of course, such ponderings bring us to river otters. In December 2015, while on the third of a four-day nature-infused writing retreat to Sapelo Island, my wife, Ruth, and I spent nearly an hour tracking a river otter along a long stretch of beach there. Had I read about river otters and their tracks before then? Yes. Had I watched video footage of river otters? Yes. Had I seen and identified their tracks before then, written about them, and seen river otters in the wild for myself, including on Sapelo Island? Yes, yes, and yes.

Yet this was different. After spotting the tracks on the south end of a long stretch of Cabretta Beach, I at first thought they would be ordinary. Granted, finding otter tracks is always a joy, especially when they are freshly made on stream banks in the middle of Atlanta, Georgia. (Seriously, folks: river otters live in the middle of Atlanta. How cool is that?) And because Sapelo has fewer than a hundred humans on it and is relatively undeveloped, the chances of coming across otter tracks on one of the beaches there is not like winning a lottery. But, still, it was a gift, and one I gladly accepted.

River otters have easily identifiable footprints, with rear feet larger than their front feet, five toes on each foot, webbing between their toes, and claw impressions associated with those toes. They also make distinctive trackways, which is not so much related to their feet but more about their gaits. One of the most common trackway patterns otters make is what trackers call a "one-two-one." In this, one of the rear feet exceeds the front foot on one side (that is one of the "one's"), but the other rear foot ends up beside that same front foot (that is the "two"), and one front foot is behind (the last "one"). The gait that produces this footfall arrangement translates to a lope, which is the baseline gait for an otter. But if an otter's second rear foot lags behind the front foot, then it is a trot, and if it exceeds the front foot (both rear feet ahead of both front feet), that is a gallop. A gallop pattern for a river otter makes a group of four tracks separated by spaces with no tracks. The four tracks then describe either a *Z* or *C* pattern, with each of those letters easily visualized by drawing an imaginary line from one track to another.[1] On the other hand (or foot), the spaces between each set of tracks represent "hang time," when the otter was suspended above the ground between each instance when its feet touched the ground. Otters also slow down enough to just walk, making right-left-right diagonal patterns in which their rear feet register behind or on the front-foot impressions on the same side.

What made these otter tracks different from others I had seen, though, was that they went on, and on, and on, showing variations of all four track-way patterns and more. These tracks spoke for the otter, telling us in no uncertain terms that walking, trotting, loping, and galloping on a beach was the only agenda on its schedule that morning. For about a kilometer (0.6 miles), Ruth and I followed its tracks northward in the sandy strip of land between the high-tide line on our right and low coastal dunes on our left. The tracks were normally only a few meters away from high tide but sometimes turned that way and vanished, then reappeared farther down the beach. These missing tracks informed us that the otter was out near peak tide that morning (between six and eight), and it was apparently cross-training, mixing up its on-land exercise regime by occasionally dipping into the surf. Raindrop impressions on top of the tracks confirmed our estimate of the timing, as the tracks looked crisp and fresh except where pitted. For us, rain started more inland that morning, moving west and south from our position around ten o'clock, but it had reached the tracks earlier. We were thus there about three hours after the rain, meaning the otter was likely long gone and on to other adventures. Nevertheless, we made sure to look up and ahead frequently, just in case the track maker decided to come back to the scene.

For those of you intrigued by animal tracks—and why would you not be?—I suggest trying to follow those made by one animal and follow it for as long as you can. That way you can learn much more about it as an individual rather than think you know it just because you can recite its species name and look up its Wikipedia page. In my experience after tracking an animal for a few hours, I can pick up on nuances of its behavior and decisions and even gain an awareness of its unique personality. For example, this otter was mostly loping but once in a while slowed to a walk or trot, or sped up when it loped or galloped, and turned to its left, right, or went straight. In short, its tracks showed enough variations to say the otter was reacting to stimuli in its surroundings, and in many different ways, leading to more questions. What gave it a reason to slow down? What impelled it to move faster? Why did it jump into the surf when it did, and why did it come out? Or do otters just want to have fun?

Admittedly, I realize that discerning a "personality" and "moods" of a nonhuman animal based on a series of its tracks might sound a little too "New Agey" for my skeptical scientist friends to accept. I also would not be surprised if those same friends made jokes about my becoming a pet

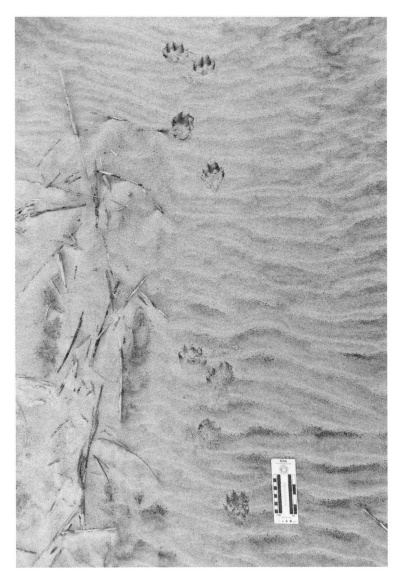

FIGURE 31. A river otter goes for an early morning stroll along a beach on Sapelo Island. *Left*, view of its tracks heading north, stepping on wind ripples just above the high-tide mark. *Right*, typical four-by-four pattern left by this otter as it galloped. Scale in centimeters (on left of scale bar).

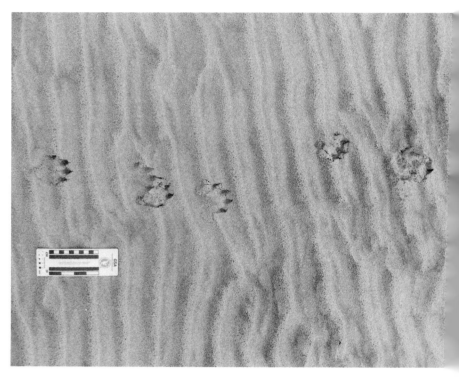

FIGURE 32. Otters are not trotters, and one of the ways to tell this is by this track pattern one-two-one, with the right rear foot to the right, the left rear and right front next to each other, and the left front foot left behind. This means its gait was a lope rather than a walk, amble, or gallop. Scale in centimeters.

psychic. (If the pay is right and it makes people happy, would you blame me?) As a fellow skeptical scientist, I am totally fine with them doing this. In fact, I will gleefully join them in making light of people who tell us that, say, they know what a Sasquatch was thinking as he strolled through a forest while somehow successfully avoiding all cameras and other means of physical detection.

But here is what happens when you have tracked a lot (which I have), made lots of mistakes while tracking (ditto), and later corrected them (I hope so). Intuition kicks in, and it usually works. For instance, at one point in following this otter, Ruth and I lost its tracks on a patch of hard-packed sand. This could have been avoided by my getting down on my hands and knees to look closer, but I was being lazy, because, you know, writer's retreat. So we put ourselves in the place of the otter and asked, "Where would I (the otter) have gone?" and looked about ten meters (more than thirty feet) ahead in what felt like the right place. There they were: the familiar one-two-one pattern, made by the same individual otter. This ichnologically inspired clairvoyance succeeded three more times, results that led me to conclude this was almost like some repeatable, testable, falsifiable, science-like thing happening. Of course, it was, as part of this predictive method was related to noticing how the tracks stayed in subtly low-elevation areas of the beach. From the otter's perspective these places were little valleys between slightly higher sandy ripples and cordgrass hillocks. A little bit of animal empathy goes a long way, as did the otter.

How does all of this relate to writing about a place? Because of that otter and its tracks, I now understand at least one river otter and its territory much better than before, and I feel like I can write with a little more authority about otters in general, like what you just read. So now you too understand this otter and its home a little better now, thanks to it leaving so many tracks while enjoying a morning at the beach. For that, I am grateful to that otter and what it taught me and, indirectly, you.

Part IV

The Human Touch

18

Alien Invaders of the Georgia Coast

Paleontologists face enough of a challenge as we try learning how parts of modern environments—such as those of the Georgia coast—might translate into the geologic record. First, we must consider taphonomy, acknowledging that almost none of the living and dead bodies we see in any given environment will become fossilized. We also admit that relatively few tracks, trails, burrows, or other traces made by animals will become trace fossils. But these are more likely than body parts to survive the present, especially if their makers lack shells, bones, and teeth. So at least we can count on traces.

Because of this pessimistic (but realistic) outlook, paleontologists often rub a big mental eraser across whatever we draw from a modern ecosystem, telling ourselves what will *not* be there millions of years from now. We then retroactively apply this taphonomically inspired actualism to what happened thousands or millions of years ago, using the assumption that today's processes provide a small window through which we can peer, giving insights into the history of the prehuman past.

Nonetheless, there is a huge complication in our quest for actualism that ecologists face every day, one the general public and maybe even nature enthusiasts either miss or ignore. The ecologists' dilemma is that the communities of nearly every ecosystem on this planet are mixtures of native and alien species, with the latter introduced—intentionally or not—by us. Thus, when we watch modern species behaving in the context of their environments and leaving evidence of their effects and evolutionary histories, we must also ask ourselves how nonnative species have cracked the window through which we darkly squint into the past.

This theme of nature altered by humankind was partly considered in Charles C. Mann's book *1493: Uncovering the New World Columbus Created.*[1]

His main consideration was how human interactions connected to trade, slavery, and genocide rapidly changed our world. But he also pointed out how nearly all terrestrial ecosystems were permanently altered by the rapid introduction of exotic plants and animals worldwide following Columbus's landfall in the Western Hemisphere. The study of invasive species and their effects, focusing on the ecological changes of the past few hundred years, now warrants its own subdiscipline—invasion ecology—one deserving of entire books of its own.[2]

Paleontologists and ecologists do not just ponder (or curse) post-Columbus alterations of actualism and ecosystems, respectively. They also think in more geologic timescales about the influences of humans on landscapes, including those of the Georgia coast. For instance, not long after humans began living on the Georgia barrier islands about five thousand years ago, people of Southeast Asian descent floated on boats to mainland Australia and brought wild dogs with them. These canines, now called "dingoes" (*Canis lupus dingo*), then spread throughout the continent.[3] Enough time elapsed since their arrival that many people in Australia consider dingoes "native," including indigenous people who include them in dreamtime stories.[4] Still, the sudden introduction of this apex predator irrevocably changed the terrestrial environments of an entire continent, with the dingoes' arrival linked to the extinction of the largest native carnivorous marsupials in mainland Australia, thylacines (*Thylacinus cynocephalus*) and Tasmanian devils (*Sarcophilus harrisii*).[5]

Examples like this show how European colonization and its aftermath in human history during the past five-hundred-plus years in the Americas were not the sole factors in the spread of nonnative species, and hint at how species invasions have been an integral part of humanity moving throughout the world. In fact, if we hold a mirror up to ourselves as a species and accept that our native homeland is Africa, we might also consider how a fire-wielding upright primate with scientific reasoning (particularly tracking abilities) and incredible cooperative skills might have been less than benign when it moved to new continents with ecosystems unused to its ways.[6] To put it bluntly, we are an invasive species too.

This self-realized perspective implies that well-meaning descriptions of "pristine," "untouched," and "unspoiled" applied to the Georgia barrier islands are at best misguided and at worst deluded. As beautiful as these places might seem (or are), we cannot view them as ecological havens immune to the effects of alien invaders, including humans who occupied some islands thousands of years ago. Moreover, like many barrier-island systems world-wide, the islands differ greatly in invader species, numbers of individuals of each species, longevity of those species, their number of introductions (accidental and purposeful), and relative degrees of how these organisms affect island ecosystems.[7]

This is one of the reasons why I devote the next four chapters to the traces of invasive species—tracks, trails, burrows, and so on—despite their makers failing an "ecological purity test" for those who might prefer focusing on native species and their traces. With regard to invasive species, the genie is out of the bottle, so we might as well study what is there rather than apply yet another metaphorical eraser to species that are drastically shaping modern ecosystems and affecting the behavior of native species, thus likewise altering their traces.

What are some of these invasive species? What distinguishes an invasive species from a merely exotic one? How do the traces of invasive species affect native species on the Georgia barrier islands and the ecology and geology of the islands themselves? And how do paleontologists and geologists contribute to our study of invasive species, including us? These are all questions that I hope to explore. But for the sake of simplicity, and before getting to the ultimate anthropocentric species that spends an inordinate amount of time pondering itself, I will showcase three invasive species of terrestrial mammals and their traces first, then an insect and its fungal cohort. The mammals are large and charismatic, consisting of feral cattle (*Bos taurus*), horses (*Equus caballus*), and hogs (*Sus scrofa*). Moreover, they are impressive tracemakers, leaving little doubt of their effects on their landscapes. The redbay ambrosia beetle (*Xyleborus glabratus*) and its fungal partner in invasive crime, laurel wilt (*Raffaelea lauricola*), are also highlighted for showing how the synergistic effects of two nearly invisible species can change the character of entire maritime forests.

Yes, I am well aware of the other invasive species on the Georgia coast, including feral cats, which deserve an entire book of their own (but one I'm never writing), and marine species, such as European green crabs (*Carcinus maenas*) and lionfish (*Pterois volitans*).[8] Nine-banded armadillos (*Dasypus novemcinctus*) are another interesting example, as these unusual mammals traveled from Central America in the nineteenth century, then made their way to the southeastern United States and the Georgia islands through a

FIGURE 33. A recently arrived and unwanted "guest" on the Georgia
barrier islands, the nine-banded armadillo, digging wherever it goes and
hence altering environments. *Left*, armadillo peeking up from a stand of
sea-oxeye daisy near a Sapelo Island beach. *Right*, a burrow dug by this
same armadillo into a sandy spot underneath a pile of driftwood, its own
beachside home.

combination of human assistance and their own means.[9] But I had to be choosy to make a point; hence the species covered here are among the most ecologically significant species in marginal-marine and terrestrial environments that also make traces inherently obvious to even the most casual observer. These are the species that I think best teach us how the Georgia barrier islands serve as unnatural laboratories that help us better understand the effects of invasive species worldwide and how invasive species are actualism game-changers for geologists and paleontologists.

With those nonhuman case studies in mind, the concept of humans as the ultimate tracemakers on the Georgia coast will also be considered, starting with Native Americans, then moving to European colonization and slavery, and on to the present with its rapid sea-level rise, increased numbers and ferocity of hurricanes, and other effects of climate change. We will then think about how these traces of ecological change translate into geologic change, with the visible evidence for change perhaps outlasting us as a species as we move toward an unprecedented future for the Georgia coast and, for that matter, every other coast.

19

The Wild Cattle of Sapelo

No one expected a bull on the beach. Yet there were his tracks, big concavities crossing a back-dune meadow, traveling up and over the dunes, then turning left, pockmarking the berm just above the high-tide mark. As we followed his hoofprints down the long stretch of beach, they occasionally vanished, denoting where he wandered into the surf. Because the tide had dropped since, this absence was in itself evidence, telling us he was there only about four or five hours before. We even found a place where he decided to drop a voluminous pile of digested vegetation on top of his trackway. Ironically, this deposit served as a reminder that any doubt of his presence could be countered truthfully by pointing at it and saying, "Bullshit." It would have been a wondrous sight, this massive red-brown-black Brahman emerging from behind a dune, strolling down the beach, walking in the surf, and owning that environment as the largest mammal on the island since the end of the Pleistocene epoch.[1] Yet the reason why it was there on that beach was because no humans were present to witness it: just like most of the Pleistocene.

I had been to Sapelo Island dozens of times before then, but this discovery of bull tracks on one of its beaches once again confirmed how my keeping an open mind would be rewarded with a completely new and delightful observation. In this instance I was with a few colleagues from Emory University in September 2015 for a three-day stay. The purpose of our trip was to visit different parts of the island and film short segments about natural and human histories for the *Georgia Coast Atlas* project. Most of these bits featured me as a "talking head" (but, unlike some gulls, still embodied) supplemented by aerial-drone footage of Sapelo's spectacular environments. The videos were later edited into informative and digestible clips shared publicly on the

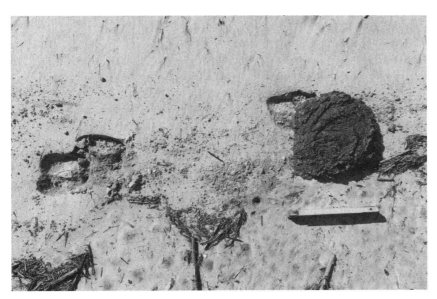

FIGURE 34. Feral cattle deciding to visit the beach on Sapelo Island. *Left*, deep, impactful feral bull tracks just above the high-tide mark, moving toward the viewer. *Right*, tracks with indirect register (rear foot registering on the front) with a metaphor deposited on the next set of tracks. Scale about twenty centimeters (eight inches) long.

atlas.[2] It was on our second day on Sapelo, after just finishing our fieldwork for the day and walking back along Nannygoat Beach, when I saw the anomalous tracks of a beach bovine.

We had some warning of the bull's nearby presence earlier the same day while filming a short video clip in the maritime forest. In the video I am squatting next to a fecal pile and tracks left by this same bull, talking excitedly about the feral cattle of Sapelo Island.[3] I also mentioned how, despite their sizes and numbers, we rarely see these animals; hence we must rely on tracks, scat, and other traces to verify their presence. Our finding much fresher bull tracks later that day on the beach accordingly felt taunting, as if their maker was daring us to find him.

So I was delighted the following day when our two-person film crew—Anandi Salinas Knuppel and Steve Bransford—proved me wrong by actually spotting the bull. He was lurking in part of the maritime forest near the University of Georgia Marine Institute, shielding himself partly from view by standing behind a clump of saw palmetto (*Serenoa repens*). My colleagues wisely used the zoom lens on their video camera to get footage of this massive, red-brown, and intimidating Brahman. A broken horn on his left side added to his ominous mystique, an anatomical alteration speaking of a past violent encounter with another male of his kind. My colleagues stayed long enough to get a few minutes of footage, but his bulk and broken horn, together with him pawing the ground while snorting at the camera crew, gave them sufficient reason to stop filming and leave the scene, posthaste.

The feral cattle of Sapelo Island are big reminders of how all of our facts and feelings about the island as some idyllic and ecologically pure place and as a true slice of what a Georgia barrier island should aspire to be are wrong. For one, people have ecologically molded Sapelo for a long time (by human standards, that is). Starting with Native Americans about 4,500–5,000 years ago, Sapelo and many other Georgia barrier islands have been transformed substantially because of large-scale changes in habitats, as well as through the introduction of exotic plants and animals during colonial and postcolonial times.[4]

Of the exotic animals introduced to Sapelo that have endured, the most significant are its feral cattle. It is the only truly wild population of cattle on any barrier island in the United States and one of the few anywhere in North America.[5] These cattle are not only free-range but also simply free, liberated from any human supervision or connection and avoiding their traditional breeders and captors whenever possible.

Unlike the feral horses on Cumberland Island (discussed in the next chapter), nearly everyone agrees on the origin story of the feral cattle on Sapelo. They most likely descended from domestic cattle released on the island by millionaire R.J. Reynolds Jr., who through wealth and intimidation tactics acquired much of the island's land from 1933 until his death in 1964.[6] Although details are sketchy as to exactly when and why, at some point during Reynolds's time on Sapelo he released a population of Brahman-breed cows and bulls, which he owned for conventional dairy and meat purposes. This animal-husbandry legacy is also reflected in the external architecture of the main building of the University of Georgia Marine Institute, which still looks like a barn. Since then many generations of these cattle have bred in the wild, and to this day they wander the island in sufficient numbers to warrant attention from wildlife biologists, ecologists, and others interested in learning about their behavior and impacts on local ecosystems.

In my everyday experience, though, the words "wild" and "cattle" are rarely used in conversations about these animals that, through our domestication of them, provide us with milk, cheese, and cheeseburgers. Also, typecasting reigns supreme when you ask a typical city dweller to describe a cow, and their descriptors are not flattering. "Slow," "docile," and "stupid" are among the most common adjectives applied, occasionally accompanied by a giggling reference to the midwestern U.S. tradition of cow tipping. (Speaking as someone who grew up in Indiana, if you do not know of this practice, I hope you never do; if you do, at least discourage others from its continuation. Be kind.)

Something important to realize about these cattle, though, is that they originally descended from aurochs (*Bos primigenius*), a wild species that survived the end-Pleistocene mass extinction about eleven thousand years ago.[7] Why yes, that would be the same end-Pleistocene extinction event that wiped out mammoths, mastodons, giant ground sloths, wooly rhinoceroses, saber-toothed cats, dire wolves, and other formidable megafauna in much of Europe and Asia. Hence aurochs must have had adaptive advantages over their Pleistocene cohorts, a survival perhaps related to their preferred original ecosystems of wetland forests and swamps. (Remember that point for later with reference to Sapelo Island.) Following the mass extinction, about eight thousand years ago, people in Eurasia, Africa, and India began domesticating aurochs.[8] Through selective breeding, people came up with the present-day varieties we see of *Bos taurus*, which, genetically speaking, is considered a subspecies of *B. primigenius*. Astonishingly, aurochs lived well into historical times, having been observed by Julius Caesar and surviving as wild populations until the seventeenth century.[9]

So whenever I go to Sapelo, I am henceforth reminded of aurochs and their evolutionary heritage, because these feral cattle are cryptic creatures of the maritime forest. Yes, you read that right: they are rarely seen, and they live in the forests. These two facts alone boggle the minds of anyone who has grown up with bucolic "old Bessie in a grassy field, chewing her cud" stereotypes of domestic cattle. Indeed, a casual day-trip visitor to Sapelo will almost never see a wild cow or bull there, let alone any of several herds that stroll through the island. When humans encounter an individual bull or herd in more open, grassy areas, the espied cattle seemingly revert to Pleistocene behavior and slip into the woods, quickly concealing themselves from prying eyes.

FIGURE 35. Varied habitats of feral cattle on Sapelo Island. *Left*, a sighting of the probable one-horned maker of those tracks and other traces on that Sapelo beach, its intimidating stare reminding us of its auroch ancestry. *Right*, a feral cattle trail cutting through a salt marsh on the north end of the island, representing an unconventional pasture. Photo on left is a still, captured from video on the *Georgia Coast Atlas* website, shot by videographers Steve Bransford and Anandi Salinas Knuppel.

In short, these cattle are not slow, docile, or stupid, and they would never allow a person to get close enough to make a short-lived and ill-fated attempt to tip any of them. Aurochs were likewise well known for their ferocity. In a 1920 painting by artist Heinrich Harder (1858–1935), titled *A Long-Horned European Wild Ox* [Auroch] *Attacked by Wolves*, he paid homage to their formidable reputations by depicting a bull handily defending himself against a pack of wolves.[10] Contrast this scenario with current expectations of how domestic cattle might behave in the face of pack-hunting predators, and you will get a better sense of just how differently the wild cattle on Sapelo Island behave.

Because of their enigmatic nature, any meaningful study of these cattle and their ecological effects on Sapelo requires the use of—you guessed it—ichnology. Accordingly, I have tracked these cattle, sometimes with students and colleagues and sometimes by myself. These tracking forays have generated many anecdotal yarns about these "wild cows of mystery" worth retelling, such as the previously mentioned bull strolling down a beach. But, for now, I will instead reluctantly restrict myself to simply describing their traces and the effects of these traces on the landscapes of Sapelo.

Traces of feral cattle on Sapelo consist largely of their tracks, trails, or otherwise trampled areas, feces, and chew marks. In my experience the vast majority of their traces are on the northern half of Sapelo, well away from concentrations of humans living in the Hog Hammock community and the Marine Institute in the south end. But wayward herds or rogue bulls occasionally leave their marks in the southern half of the island when they graze in grassy areas or travel through forests there.

Tracks made by these feral cattle are unmistakable when compared with those of other hoofed animals on Sapelo, such as white-tailed deer (*Odocoileus virginianus*) or feral hogs, and their footprint distinctiveness is a function of both shape and size. Tracks look like robust inverted Valentine's hearts, with two bilaterally symmetrical hoof impressions rounded in the front and back. They are normally about 9–14 centimeters (3.5–5.5 inches) long, although I have seen newborn-calf tracks as small as 5–6 centimeters (2–2.3 inches) long. Track widths are slightly less than lengths, with about a 20 percent difference between the two. Like deer, these cattle spend most of their time walking slowly or standing. Hence their rear-foot impressions often overlap or register behind their front feet, but they can also overprint the front foot, which trackers call "direct register." Trackways typically show a diagonal-walking pattern, although these can be punctuated by frequent "T-stops." In these, tracks form a *T* pattern, with the top of the *T* made by the front feet placed next to one another whenever a calf, cow, or bull stopped. In at least one instance, I found two T-stop patterns made conjointly by an adult bull behind a cow, resulting in six tracks, which I hope I don't need to explain any further.[11] According to their tracks, these wild cattle rarely trot or gallop, and if they do, it is likely a result of pesky humans encroaching on their territory.

Because the wild cattle of Sapelo obey mostly herding instincts, they habitually follow one another along the same narrow pathways through maritime forests and salt marshes. These movements result in well-worn trails that wind between live oaks (*Quercus virginiana*) in forest interiors or cut straight across marshes. Nonetheless, the cattle also like to use open "freeways" of the sandy roads crisscrossing the northern half of Sapelo. This combination of clear roadways and sand makes tracking them much easier, especially after a hard rain has cleaned surfaces for the recording of fresh signs. When cattle travel on a road, they break out of single-file movements imposed by maritime forests and instead walk parallel to or just behind one another. This shift in their herding behavior is indicated by overlapping and side-by-side trackways. When on forest trails, they also often drag their hooves across the tops of logs downed along trails, chipping and otherwise breaking down the wood.[12]

The most interesting behavior I have seen from these wild cattle was just mentioned: making trails that cut across salt marshes. Why do they do this? It turns out that instead of restricting themselves to a terrestrial-only diet, they are eating smooth cordgrass. These forays into all-natural, locally grown, and all-you-can-eat salad bars result in the cattle leaving many other traces, such as near-ground-level cropping of *Spartina* accompanied by considerable trampling of grazed areas. I was surprised to discover this for myself in 2005, but people who raised cattle on the island in the nineteenth and early twentieth centuries—perhaps through necessity—knew about this alternative foodstuff and fed it to cattle as a substitute for hay.[13]

Of course, among the most obvious traces these cattle leave in their wake are their end products of digestion. These "cow patties" vary in size, depending on both the size of the tracemaker and the liquid content of the scat. The bigger the tracemaker and the greater the water content to the plants, the flatter and wider the patties, which can easily exceed dinner-plate diameters.

Similar to the situation on Cumberland Island with its feral horses and their feces, native dung beetles must not be able to keep up with such a bounty. As a result, I've seen many dried and nonrecycled patties throughout the island, as well as freshly unloaded pies that showed no signs of discovery by caring dung-beetle mothers.

How do feral cattle relate to geology and paleontology? Part of our consideration of them as an exotic species is that they reflect an evolutionary lineage that interacted with native environments in the past. The closest Pleistocene ecological analogue to these cattle either on or near the barrier islands then would have been bison (*Bison bison*).[14] Yet modern bison are not known to graze in coastal ecosystems. Did bison do so in the prehuman past of the Georgia coast? Perhaps, and it is a possibility made more likely when considering their great numbers before human predation decimated their populations.[15] Given what we have learned from these cattle, we now have an idea of what sort of trace-fossil evidence would tell us of a former bison presence in coastal environments.

A tantalizing hint of what this evidence might look like came from St. Catherines Island, where I once saw an exposure of relict marsh with cattle tracks impressed onto its top, made by cattle there sometime in the first half of the twentieth century. This is an excellent example of how paleontologists looking into the fossil record of aurochs or their ancestral species—whether of body fossils or trace fossils—might use present-day cattle traces as clues when prospecting for trace fossils. It also shows why I urge paleontologists to pay attention to invasion ecology and conservation biology, in which "ecologically impure" invasive species give us valuable insights on their evolutionary histories.

Perhaps the biggest difference between the feral cattle on Sapelo and exotic hoofed species on other Georgia-coast ecosystems is their impact on salt marshes there, mostly in just the northern half of the island. In all of my years of noting cattle tracks and other signs on Sapelo, I have only twice seen evidence of their going to back-dune meadows, coastal dunes, or beaches. Instead, they prefer to stay in the maritime forests and wetlands. This preference may reflect how the cattle switched back to auroch-like behaviors of the Pleistocene within just a few generations, choosing to live in wooded wetlands instead of terrestrial grasslands enforced by modern humans.[16] The northern half of Sapelo also has extensive salt marshes flanking maritime forests and freshwater sloughs containing other wetland plants.

Again, it makes good ecological sense that the cattle would stay mostly in that half of the island. These ecologically and behaviorally driven preferences thus place the feral cattle in the same invasive-species category as feral horses—which negatively affect coastal dunes and salt marshes—as well as feral hogs, which go even into the intertidal zones of beaches for their foraging. Other than occasional deer hunters, Department of Natural Resources personnel, or wayward ichnologists, the absence of humans on the north end of the island is also a plus, as these cattle evade people whenever possible.

What else can we learn about these feral cattle and their ecological and geologic impacts on Sapelo, especially through studies of their traces? For one, knowing the actual number of cattle on the island would be useful, as their quantity surely relates to how well island ecosystems can handle present and future populations. This information would also test whether or not the feral cattle of Sapelo Island are truly invasive or if they are just exotic. The difference between the two labels is that invasive species degrade ecosystems, whereas exotic species may be neutral in their effects or even beneficial. In short, more science is needed before metaphorically branding these cattle.

An important facet of this knowing, then, is to better define wild-cattle behaviors in the context of their nonnative ecosystems. How can we do this with a species that stays hidden so well and one that apparently reverted to an evolutionarily embedded mode of life? Fortunately, behaviors can be defined and documented through the ichnological methods outlined here. These methods can then easily augment other tools routinely used by conservation biologists, such as trail cameras, aerial drones, and direct observation. Once this is done we will know much more about these wild cattle than before and will no longer have to rely on whispered legends of the mysterious untamed bovines of Sapelo Island. Regardless, there is still plenty of room for more such stories and perhaps even artwork, operas, plays, musicals, and movies. Because cattle have played such an integral role in the development of humanity, there is every reason to suppose that they and their traces will continue to intrigue us for as long as they continue to live on Sapelo.

20

Your Cumberland Island Pony, Neither Friend nor Magic

Cumberland Island is difficult to visit, even if you live in Georgia, and more so if you live in the metropolitan Atlanta area. Because it is the southernmost of the Georgia barrier islands and just north of the border with Florida, it is a minimum six-hour drive from Atlanta. How about taking a train there? Please. This is the southeastern United States, where passenger trains underwent an extinction-level event with the inception of the interstate highway system during the 1950s. Then perhaps some of that driving time could be cut by flying out of Atlanta and getting a rental car, right? But the nearest sizable airport is in Jacksonville, Florida, which is still about an hour drive away from St. Marys, the town closest to Cumberland on the Georgia coast. Once you add up the time of getting to the Atlanta airport, going through security, flight delays, waiting for checked bags at the other end (if they arrive), getting a rental car, and driving to St. Marys, your transit time may have equaled or exceeded that of driving, with the special bonus of humiliating full-body scans and other forms of security theater.

So, driving it is. If you arrive in St. Marys in the late afternoon or evening, though, you will not be able to get onto the island until the next day. Cumberland has no bridge linking it to the mainland and no airport, and I would not recommend swimming there. That means you must take a boat, which, if you have one handy, great, but most people don't. Fortunately, you can purchase a ticket from the U.S. National Park Service office at the dock in St. Marys the next morning and take a ferry from there to Cumberland— that is, unless they sold out in advance, because the Park Service has decreed a three-hundred-visitor limit to the island per day.[1] If you are lucky enough to get a ticket, the ferry ride to the island then takes a little less than an hour. Granted, it is a pleasing trip on a tidal channel that winds past salt marshes, offering glimpses of bald eagles, white and brown pelicans, and much more. Yet this hour on the boat means one less hour on the island itself, particularly if you are coming back the same day. Staying the night on Cumberland normally offers one of two options: camping in a campground that requires a reservation well in advance or spending thousands of dollars in a luxury resort.

I've visited Cumberland perhaps a dozen times, normally while leading students on field trips (such as when we learned about coquina clams together) but also while looking for crayfish burrows and other animal traces in its interior. On those visits my wife, Ruth, accompanied me, and we pitched a tent on the property of Cumberland's resident expert naturalist, Carol Ruckdeschel. Carol, who has lived on Cumberland for more than four decades, has been an invaluable teacher and colleague whenever we've done fieldwork there, taking us to places most of the public would never see and otherwise helping us better learn about Cumberland's remarkable ecological nooks and crannies.[2] Compared with most people, then, I consider myself a Cumberland veteran, yet I still feel like a dilettante compared with a scientist as experienced and knowledgeable as Carol.

Regardless, during any given ferry ride to Cumberland, I can't help but feel like a seasoned explorer (and bordering on judgmental) as I listen to the excited chatter of first-time visitors. Why such a high-and-mighty disparaging attitude? Because in nearly all such conversations, these day-trippers talk about what they just read about Cumberland Island before boarding the ferry, and I guarantee they do not mention its coquina clams, ghost crabs, ghost shrimp, periwinkles, insects, crayfish, horseshoe crabs, or other invertebrate tracemakers. Instead of horseshoe crabs, they talk about shoeless horses. Inevitably, one person will state the following sentiment with an authoritative air while gazing at Cumberland: "Ah, yes, the wild horses of Cumberland Island. Roaming free since the time of the Spanish in a pristine, unspoiled landscape, grazing contently on the sea oats and strolling through the coastal dunes, in perfect harmony with nature."

Yet everything in that statement other than the words "horses of Cumberland Island" is wrong. And as an educator and an author who writes for popular audiences, I feel responsible for correcting inaccurate proclamations made about any aspects of my beloved Georgia coast. So I have a sense of educational penance while trying to right the wrongs promulgated about the feral horses of Cumberland Island. As shown on Internet searches, magazine articles, book covers, and other popular media, the Cumberland

FIGURE 36. The traces of feral horses on Cumberland Island, small and large. *Left*, horse tracks crossing a modest dune field on the south end of Cumberland. And just why are those dunes so low and sparsely vegetated? *Right*, trampling and overgrazing of a salt marsh, with its smooth cordgrass reduced to mere nubs in broad swaths.

horses are without a doubt the most charismatic of nonnative animals on any of the Georgia barrier islands. Yet more than any other invasive species, they are also lightning rods for controversy and heated arguments. I have also noticed these acrimonious provocations become most acute whenever anyone tries to inject actual history or science. So, with the understanding that discussing the horses of Cumberland Island is akin to walking into a tall-grassed field filled with their droppings, I will talk about them from my unique perspective as a paleontologist and geologist. I hope this viewpoint will add another dimension to what is often presented as a two-sided and emotionally charged argument.

Among the Georgia barrier islands, Cumberland is unique because it is part of the U.S. National Park system as a National Seashore and because it is the only Georgia barrier island with a population of feral horses. Nevertheless, despite the fame of these horses—figuring as key attractions in advertisements about Cumberland and inspiring dreamy book titles—their origin stories remain murky and perhaps deliberately so. One of the recurring romanticized claims is that these horses descended from livestock brought there by Spanish expeditions in the sixteenth century.[3] This idea is reassuring to the people who repeat it for two main reasons: (1) it establishes horses as living in the landscape for a long time (especially by American standards), meaning that their presence there now is considered natural; and (2) for people of European descent, it lends a comforting connection of horses to their heritage, but imprinted on an American place.

But once said enough times, these just-so stories become faith based, and any evidence contradicting them is intolerable. Thus, even when historical accounts reveal that the horses are from a population released in the first half of the twentieth century, and genetic studies show they are not appreciably different from horse populations on other islands of the eastern United States—arguing against a purely Spanish origin—any questioning of the stated premise (in my experience) provokes angry responses from its defenders.[4] I suspect this virulent reaction is a direct result of challenging both the "naturalness" and "cultural heritage" of the horses on Cumberland. In reality, though, these are opposing values. After all, even an admission that these feral horses came from European stock at any point during the past five hundred years supports how they do not belong on Cumberland Island, or anywhere else in the Western Hemisphere, if we are talking about the past eleven thousand years or so.[5] In other words, the point is moot whether the current horse population originated in the six-teenth, seventeenth, eighteenth, nineteenth, or twentieth centuries or is a mixture of older stock and relative newcomers. If only horses could talk, we would know for sure. But lacking that, we have to use evidence-based reasoning instead.

Arguments of heritage aside, these horses are definitely newcomers in a geologic and ecological sense. The fossil record of the modern Georgia bar-rier islands backs this up, as Cumberland, Jekyll, St. Simons, Sapelo, St. Catherines, and Ossabaw Islands, with their foundations of forty-thou-sand-year-old Pleistocene sedimentary rocks, contain no bones or trace fossil evidence of horses. Granted, North America was horse-rich during the Pleistocene epoch, with more than forty species of horses (*Equus* spp.) named by paleontologists.[6] These horses, however, were all extinct by about eleven thousand years ago, and none are known to have inhabited any of the barrier islands with a Pleistocene record. Before humans, the animal most ecologically similar to horses on any of the Georgia barrier islands would have been bison, but their bones are rare in coastal deposits.[7] This scarcity leads me—and probably other paleontologists—to wonder whether the is-lands ever had self-sustaining populations of large herbivores. This is not to say that the Pleistocene barrier islands did not occasionally host giant ground sloths, mammoths, or other large herbivores. But such big mammals bringing outsized appetites with them would not have lasted long after out-stripping the resources of those islands.

With all that human history and prehistory in mind, the traces made by the feral horses of Cumberland and their ecological effects are exceptional to Cumberland and every other Georgia barrier island and hence worth our attention. Just to keep this simple, I cover three primary types of horse traces: tracks and trails, chewing, and feces. What these traces all have in common (other than being made by a horse, of course) is their decidedly negative impacts on the native plants and animals of Cumberland, includ-ing ecologically iconic species of the oft-labeled "pristine" ecosystems of the island.

Tracks and trails are the most abundant and easily spotted of horse traces on Cumberland, even to someone with little or no training in ichnology. Horses are unguligrades, which means they are walking on their toenails (unguals in horses are more commonly called "hooves"), which cover a single digit. Horse hooves make circular to slightly oval compression shapes, but if preserved in a firm mud or fine sand will leave a "Pac-Man"-like form. Front-foot tracks are slightly larger than rear-foot tracks, at 11–14 centimeters (4.3–5.5 inches) long and 10–13 centimeters (4–5.1 inches) wide, whereas rear feet are 11–13 centimeters (4.3–5.1 inches) long and 9–12 centimeters (3.5–4.7 inches) wide. Variations in track size typically depend on relative ages and sexes of the horses, with the smallest made by foals and yearlings, then larger ones made by colts and mares, and the biggest belonging to stallions.[8]

An important point to keep in mind when tracking horses or any other hard-hoofed animals, such as feral cattle, feral hogs, or deer, is that their feet readily cut through sediments and vegetation. This means they form more sharply defined and deeper impressions than the padded feet of an equivalent-sized animal. Because Georgia coast sands contain whitish quartz and darker heavy minerals, these contrasting sand colors better outline horse tracks on those surfaces. In cross-sections, then, horse tracks can be discerned as deep and clearly outlined structures that cut across the bedding in a sand dune or berm.

When asked to think about horses in motion, it might be tempting to imagine them galloping, especially along a beach at sunset, perhaps ridden by a shirtless Fabio. Still, a horse would tire quickly if it galloped all day, especially for no valid reason other than fulfilling human romanticism. Its normal gait is instead a slow walk, which causes the rear foot to register partly on top of the front-foot impression and slightly behind.[9] With a faster walk the rear foot exceeds the front-foot impression. From a walk to a trot, trackways form diagonal-walking patterns, as the right-left-right alternation of steps can be linked with imaginary diagonal lines. Trackway width, or straddle, is about twenty to forty centimeters (eight to sixteen inches) if a horse walked normally but narrows noticeably if the animal picked up speed to a trot, lope, or gallop.

Given enough back-and-forth movement by horses along preferred routes, their overlapping trackways form trails. These paths can be picked out as linear-to-meandering bare patches of exposed sand or mud cutting across ground vegetation. Because horses are much larger than native white-tailed deer on Cumberland, their trails are considerably wider and deeper. These trails are so prominent and numerous on the southern end of Cumberland that one can literally see them from space. The latest satellite photos taken of Cumberland, viewable via Google Earth, show a spiderweb network of horse trails cutting across high marshes, back-dune meadows, and dunes, trails absent on all other Georgia barrier islands.[10]

Chewing traces are obvious once you start looking for them on Cumberland, because to horses the entire island is a salad bar. Horses are grazers—eating at ground level—but also low-level browsers, eating just below or at head level. Because they eat a wide variety of vegetation on Cumberland, they inevitably consume ecologically important plants. These include smooth cordgrass, sea oats (*Uniola paniculata*), and live oak, all three of which are integral species in their respective ecosystems. As we learned earlier, smooth cordgrass predominates low salt marshes; sea oats are the mainstay plants of coastal dunes, with their root systems helping to stabilize and build dunes; and live oaks are the largest and most long-lived trees in the maritime forests, affecting nearly all other plants.[11] The negative effects of horses consuming smooth cordgrass and sea oats are straightforward, as these plants hold sediments in place and keep them from eroding; trampling of these areas also prevents growth and even decreases ribbed mussel populations.[12] But how do horses affect live oaks? They eat seedlings, and an absence of these means older oaks are being replaced by younger ones much more slowly. If the horses continue doing this, they will contribute to the degrading and fading of the maritime forests of Cumberland.

Grazing traces consist of clean cuts of vegetation within a vertical swath over a broad area. Horses have teeth on both their upper and lower jaws; thus they shear plants on their branches, stems, or leaves. In contrast, deer leave more ragged marks, as they lack front teeth on their upper jaws and hence must clamp onto and tug on vegetation to break it off. Horses also create browse lines, which are abrupt horizontal lines of decreased vegetation that more or less correspond to a range of horse-head heights.[13]

A subtler trace of horse browsing on Cumberland Island is discernable by looking for what is not there, which is Spanish moss (*Tillandsia usneoides*), one of the iconic species of the Georgia barrier islands and much of the lower coastal plain, hanging down as wispy, grayish-green, hairlike tendrils from live-oak branches. Just like the horses, though, Spanish moss is not of Spanish origin (nor is it moss). This plant is actually more closely related to pineapples, as it is a bromeliad; bromeliads are often nicknamed "air plants" because most do not need soil, gaining their nutrients from the air.[14] This adaptation is especially apparent with Spanish moss, which flows with every little breeze. On Cumberland Spanish moss is absent or rare in parts of the maritime forests there, a victim of overly consumptive horses who pull it down from the trees.

FIGURE 37. If you think the feral horses on Cumberland Island belong there, take a look at their effects on the landscapes, and you might wonder whether you're backing the wrong horse. *Left*, two horses grazing in back-dune meadows and immediately behind coastal dunes. *Right*, a well-established horse trail cutting through the edge of a maritime forest, one of hundreds on the island.

For day-trip visitors on Cumberland, horse dung is definitely the most multisensory and visually obvious trace on the landscapes there. During any given stroll, one cannot avoid seeing, smelling, and stepping in horse feces. This teeming overabundance means that horse wastes are not being recycled quickly enough into the ecosystems, taxing even the most industrious of dung beetles.[15] For sure, I have seen a few traces of dung beetles in fresh piles of horse feces, evident as small, round holes where dung-beetle mothers have gathered fresh food for their potential offspring.[16] But no matter how much I have looked or wished for it, I have yet to witness great thundering herds of beetles majestically rolling balls of dung across Cumberland Island landscapes.

One of the more interesting ecological consequences of horse dung I have seen on Cumberland is how horse pellets or piles form tiny landscapes on their own, influencing or otherwise modifying the behaviors of smaller animals. For example, on sand dunes next to Lake Whitney on Cumberland— the largest body of freshwater on any of the Georgia barrier islands—I was surprised to see how small lizards, such as skinks, moved around the dung piles or burrowed under them, using dried feces as roofs or at least as, well, shingles. But the lizards' attraction to these equine end products also may have been partly motivated by who else was visiting them, such as beetles and dung flies, thus providing the lizards with easy meals.

All three categories of traces—tracks, chews, and dung—are easily found together in ecosystems wherever horses are trampling, grazing, and defecating, respectively. Hence knowing what we do about these traces, we can then adopt our now-experienced paleontologist or geologist personas and ask ourselves about the likelihood of them preserving in the fossil record and how we would recognize them if they did.

Speaking as a paleontologist, I predict their likelihood of preservation, in order, would be trails, tracks, feces, and chews. Trails might be evident as wide, linear, and compacted depressions cutting across relict-marsh surfaces that also lack root traces and fiddler-crab burrows. Such trails may also preserve tracks, visible as large compression shapes in a horizontal bedding of mud or sand or as deep disruptions of sedimentary bedding in vertical sections.[17] Feces, or their fossil versions called coprolites, might get preserved. Still, herbivore feces filled with vegetative material are less likely to make it into the fossil record compared with carnivore feces, which are premineralized because of their bone content.[18] Chew traces would be nearly impossible to tell from normal tearing and other degradation of plant material before it fossilized, but a trained eye might detect them.

Could the widespread and long-lived ecological damage caused by an invasive species serve as a sort of trace fossil itself? In the case of horses or ecologically similar animals, subtle changes to landscapes over time might take place that then manifest themselves geologically. This experiment has already been done on Assateague Island of Virginia and Maryland, which also has a feral-horse population. In the summer of 2018, Ruth and I visited this barrier island for a day hike, and we stayed long enough for her to get chased off a trail by two aggressive stallions. (She survived.) Assateague is much longer than Cumberland, but the island has enough horses and for a similar amount of time (about a hundred years) that these large, non-native herbivores have altered its ecology and geology. Coastal geologists who studied Assateague found that areas that used fences to exclude horses had dunes that averaged 0.6 meters (2 feet) higher than those with over-grazed and trampled dunes.[19] Coastal geologists conducted a similar study on Shackleford Banks of North Carolina by examining places where fences separated horse-occupied from horse-devoid parts of the island.[20] These geologists likewise found that horses caused dunes to be less than 1.5 meters (5 feet) high, whereas dunes without horses were more than twice as high, as much as 3.5 meters (11.5 feet). Decreased dune heights mean that storm waves can more easily penetrate these natural barriers, which in turn results in more frequent and widespread storm washovers. For this and other reasons, Assateague and Cumberland, which are both managed by the U.S. National Park Service, have strict rules against people walking across dunes or pulling up sea oats. Yet the horses on both islands do both every day and all year round and far more often than clueless or malevolent tourists.

If this geologic evidence is not persuasive enough, then you also might think about shorebirds. Such changes in the coastal geology of primary dunes and back-dune areas result in fewer ground-nesting shorebirds. Smaller or absent coastal dunes translate into less protection against storm waves, which causes more frequent inundations of nests and the drowning of eggs and nestlings.[21] Horses also can have a direct impact (literally) by stepping on shorebird eggs and nests or scaring away parents from nests, with the latter increasing the chances that eggs overheat in direct sunlight, or predators take out the next potential generation of shorebirds. Again, Cumberland and other islands post signs warning people not to walk across back-dune areas during shorebird nesting seasons for exactly these reasons. But, alas, horses cannot read, nor are they likely to learn.

With all this said, I fully expect naysayers will still want to fight me, and I predict their opening salvo will be pointing out that horses were native to the Americas, so we're just returning what rightfully belongs there, as a sort of equine manifest destiny. So let's do that mental exercise and imagine mainland horses making it to a Georgia barrier island during the Pleistocene epoch and establishing a breeding population. The geologic sequence following their arrival would look like this, from bottom to top: high dunes suffused with root traces from sea oats and other plants (before horses); lower dunes corresponding with fewer root traces and deep disruptions of dune bedding (horse tracks); and increased frequency of storm-washover fan deposits that might be succeeded vertically by either a salt marsh or maritime forest. Similarly, a paleontologist would find dramatic decreases in root trace fossils; ghost-crab, insect, mouse, and mole burrows; and shorebird nests, eggshells, and tracks, possibly culminating in a local extinction of shorebirds. To summarize, a paleoecologist would detect an overall progression from a dune-dominated shoreline with abundant evidence of high biodiversity to a completely different ecosystem with noticeably lower biodiversity. Through trace fossils and modern traces, the past and present can be used to predict the future.

From a geologic and paleontological perspective, should the feral horses of Cumberland Island be removed? Yes. Will they be removed? Probably not. As much as I admire the personnel of the U.S. National Park Service, they know which way their bread is buttered, and great numbers of people coming to see the horses enhance the local economy. So it is not surprising to see how online and print advertising for Cumberland Island and Assateague Island prominently feature their wild horses, which attract hundreds of thousands of annual visitors who flock to see this livestock, which is somehow made more glamorous when walking, feeding, and defecating on a barrier island.

Regardless of what happens, though, I will keep teaching about the horses of Cumberland Island and their traces, both as an educator and a concerned citizen. Perhaps with enough awareness, circumstances will change for the better so that Cumberland Island can not only remain a beautiful place but also heal and become more like what it was before the arrival of horses there, which was not so long ago.

21

Going Hog Wild
on the Georgia Coast

Anytime I hear someone refer to a Georgia barrier island as "pristine," I wince, give them a little smile, and say, "Well, bless your heart." As you know by now, nearly every island on the Georgia coast—no matter how picturesque—is not in a primeval or unsullied state, having been considerably altered by Native Americans, Europeans, Africans, and Americans over a minimum of the past four thousand to five thousand years. The varying degrees of change are sometimes subtle but nonetheless present, denoted largely by the loss of original habitats and native species or the addition of nonnative species.

Still, one Georgia barrier island comes close to fulfilling this idealistic description: Wassaw. Part of this uniqueness is because Wassaw is a geologically young island, having formed only a little more than a thousand years ago.[1] This did not leave much time for ecological succession to birth and nurture its maritime forests, freshwater wetlands, marshes, and dunes, nor for Native Americans to occupy and modify these places. Even after European colonizers arrived, the island escaped commercial logging, agriculture, animal husbandry, and extensive year-round settlements. Also, unlike every other Georgia barrier island, an African American owned it for a while. Anthony Odingsell inherited Wassaw from his former master and lived there during much of the nineteenth century; only one family (the Parsons) possessed it afterward.[2] There they placed a house in its center and made a network of sandy roads, but otherwise let Wassaw be Wassaw. But it was modified greatly by the 1893 Sea Islands Hurricane, which inundated the island and all others on the coast.[3] An eerie reminder of that hurricane and its power remains as a thick deposit of sand dumped by storm waves then, evident today as a tall ridge in the middle of a maritime forest and far from the present-day shore. In 1969 the Parsons sold the island to the

Nature Conservancy, which turned it into a U.S. National Wildlife Refuge, managed by the U.S. Fish and Wildlife Service.[4] Today it functions as a nature reserve, especially for ground-nesting shorebirds and sea turtles, both of which nest seasonally there. It is also one of the few Georgia barrier islands that alligators have reclaimed as its original (and rightful) rulers, with wallow-initiated ponds and other freshwater wetlands punctuating lowland areas in maritime forests and elsewhere.

Given all this naturalness, one might surmise this is why Wassaw works so well as a refuge for birds and sea turtles. Yes, this is true, but it is also because of Wassaw's status as a relatively hog-free island. I have been lucky enough to visit Wassaw five times, the first on a geology field trip in 2007, in which the trip leaders compared and contrasted Wassaw with its hyper-developed neighbor to its north, Tybee. Four successive visits there were with my students in tow, all led by John "Crawfish" Crawford, who was mentioned earlier with reference to a decapitated seagull that was totally not his fault. In all those admittedly brief and infrequent visits, I saw no tracks or other traces of hogs there, and I remain hopeful that the island will continue to enjoy a hogless future.

Feral hogs have a special place in the rogue's gallery of invasive mammals on the Georgia barrier islands, and all but the most recalcitrant of pro-porcine contrarians agree that they are the worst of the lot. Hogs are on every large undeveloped island—Cumberland, Sapelo, St. Catherines, and Ossabaw—and they bring ecological mayhem with them wherever they roam. The widespread damage they cause is mostly related to their voracious omnivorous diets, in which they seek out and eat nearly any foodstuff, whether fungal, plant, or animal, alive or dead.[5] Their fine sense of smell is their greatest asset in this respect. Every time I have tracked feral hogs, their tracks show head-down and nose-to-the-ground movement as the norm, punctuated by digging that uses a combination of their snouts and front hooves to tear up the ground in their quest for food, which upon finding they gluttonously gobble. In short, they act like, well, you know.

Most important, from the standpoint of animals native to the Georgia barrier islands that try to live more than one generation beyond a single hog meal, feral hogs eat eggs. Hence ground-nesting birds and turtles are their most adversely affected victims, and hogs are particularly keen on eating sea-turtle eggs.[6] Mothers of all three species of sea turtles that nest on the Georgia coast—loggerhead (*Caretta caretta*), green (*Chelonia mydas*), and leatherback (*Dermochelys coriacea*)—dig subsurface nests filled with 100–150 eggs full of protein and other nutrients, making them tempting targets. Yes, native species, such as ghost crabs and raccoons, also eat sea-turtle eggs. But ghost crabs—which burrow down into the egg chambers—normally nosh on only a few eggs at a time, resulting in small losses to a clutch of more than a hundred eggs.[7] By digging down to an egg chamber, though, they expose its scent to any other egg predators in the area, such as raccoons. Worse than ghost crabs, raccoons either chow down on an egg clutch over several nights or invite over family or friends to dine with them.[8] Raccoons are also quite clever at finding nests. For example, on St. Catherines Island I once found raccoon tracks dead center in a mother loggerhead trackway on a beach, showing how the raccoon followed the mother's scent to her nest. But raccoons are not nearly as greedy as hogs, which eat entire nests as a turtle-egg buffet. These egg predators similarly threaten the future continuation of another saltwater turtle, the diamondback terrapin (*Malaclemys terrapin*). This turtle lays its eggs in shallow nests near the edges of salt marshes, which raccoons and hogs also manage to find. Conservation efforts, regulations, and education have halted the decline of diamondback terrapins, which once suffered from human predation (they were used as a tasty ingredient in soups).[9] But like horses, hogs cannot read, nor are they affected by human rules; hence they do not discriminate: eggs are eggs.

This is how feral hogs are particularly dangerous as an invasive species. Unlike feral horses or cattle that "merely" degrade parts of their ecosystems, hogs interfere with the reproductive cycles of some native species, which can directly lead to their extinction. Indeed, the dodo (*Raphus cucullatus*), a flightless bird native to the island Mauritius that is often used as a symbol of modern extinctions, was not a victim of excessive stupidity and naivety in the face of intellectually superior Europeans who hunted down every single bird.[10] Instead, its quick demise is now blamed on the rapid habitat destruction of Mauritius, which conspired with egg-eating hogs, rats, and goats introduced by the same Europeans.[11] As I often tell my students, if you want to cause a species to go extinct, stop it from reproducing, and hogs practice this principle all too effectively on animals that lay eggs on or in the ground.

Dietary flexibility and species endangerment aside, as an ichnologist I am most astounded by the extremely wide ecological range of hog traces. I have seen their tracks—often made by groups of animals traveling together—in the deepest interiors of maritime forests and freshwater wetlands on islands, as well as crossing back-dune meadows, high salt marshes, coastal dunes, berms, and beaches. If their traces became trace fossils, paleontologists would refer to them as a facies-crossing species, in which facies (think "face") comprise the identifiable traits of a sedimentary environment preserved in the geologic record.[12] As shown by their tracks and signs, hogs are omnipresent in terrestrial and marginal-marine environments. Moreover, they are good swimmers. They can easily swim across tidal channels at low tide, enabling them to spread from island to island without the assistance of humans.[13] As a result, a so-far hog-free Wassaw Island is under constant threat of hogs crossing over from its southern neighbor, Ossabaw Island, which is overrun by them.

So to better understand why feral hogs are such successful invaders of the Georgia islands, it helps to first think about their evolutionary history. As expected, this history is complicated, just like that of any domesticated species in which selective breeding narrowed the genetic diversity we see today. Today *Sus scrofa* has about fifteen subspecies, making its recent family tree look rather bushy.[14] As seen in genetic studies, divergence between wild species and various subspecies of hogs may have happened as long as five hundred thousand years ago in Eurasia. But humans did not capture and start breeding these animals until about nine thousand to ten thousand years ago.

The closest extant native relatives to feral hogs in North America are collared peccaries (*Pecari tajacu*), which live in the southwestern United States, Central America, and South America.[15] Peccaries, though, are recent migrants to North America from points south, and the only two Pleistocene species known from the fossil record of the southeastern United States are the flat-headed peccary (*Platygonus compressus*) and the long-nosed peccary (*Mylohyus nasutus*).[16] Furthermore, although peccaries share a common ancestor with hogs and their relatives, they evolved separately in the Americas. Hence they differ ecologically from feral hogs, particularly in their mostly plant-based diets. All this means that the post-Pleistocene ecosystems of the southeastern United States, and especially those of the Georgia barrier islands, have never encountered anything like feral hogs.

Unlike the feral horses of Cumberland Island and the feral cattle of Sapelo Island, the feral hogs of the Georgia barrier islands were introduced early in European colonization of the coast, probably starting with the Spanish in the sixteenth century.[17] In this scenario these colonizers created "living larders" by dropping off a few breeding animals on islands, which would have eaten anything and reproduced in sufficient numbers on their own. This practice assured sailors of plentiful meat supplies whenever they came back to those islands. Evidence for this origin story remains in the genes of some feral hogs on Ossabaw Island, which firmly reflect an Iberian origin, a criterion used to identify them as Ossabaw Island hogs.[18]

The living-larder concept is an unfortunate consequence of selective breeding of Eurasian hogs, in which breeders chose for early sexual maturity and large litter sizes. For instance, female feral hogs can reach breeding age at five months, and litters typically have four to eight piglets but can be greater than twelve; females also can produce three litters in just more than a year.[19] Do the math, and that's a lot of bacon making in very short amounts of time. Moreover, the only predation pressures piglets face daily on Georgia barrier islands with few year-round human residents are from raptors or alligators. This situation means young hogs reach sexual maturity soon enough to rapidly spread their children throughout a barrier island.

Yet, as we have learned in North America, and particularly on the Georgia barrier islands, feral hogs rapidly revert to their Pleistocene roots. Similar to the feral cattle of Sapelo Island, people rarely see these hogs, especially on islands where humans regularly hunt them. Nearly every time I have spotted them on Cumberland, Sapelo, St. Catherines, or Ossabaw, they instantly turn around, briefly flash their potential pork loins and ham hocks at me, and flee. Also, as anyone who has raised or hunted hogs can tell you, pigs are smart and quick learners.[20] So I imagine that after only one or two shootings of their siblings or parents, they readily associate upright bipeds with imminent death, especially if said bipeds are carrying "boomsticks." (Speaking of which, I knew of at least one sea-turtle researcher who did his part to decrease feral-hog populations through his able use of such baby-sea-turtle-protection devices, while also generously feeding local vultures.)

Because direct observations of these free-ranging pigs and their behaviors are so challenging, much of what we learn about them in the context of the Georgia barrier islands is from their traces. Among the most commonly encountered feral-hog traces are tracks, rooting pits, wallows, and feces. I will now describe each of these so that you can better detect their presence on a Georgia barrier island, or anywhere else, for that matter.

Feral-hog tracks are potentially confused with deer tracks, as they both consist of paired hoofprints and overlap in size, at about 2.5–6 centimeters (1–2.5 inches) long. Nonetheless, feral-hog tracks are more rounded on

FIGURE 38. The effects of unwanted visitors (feral hogs) in salt marshes. *Left*, extensive and deep rooting pits on St. Catherines Island that, once connected, convert salt marshes into mud flats. *Right*, a parent hog with two of its offspring in a salt marsh on St. Catherines Island, exposing their meaty parts as they flee.

their ends while having nearly equal widths and lengths, and their hooves often splay.[21] Two dewclaws (vestigial toes) also frequently register behind hog hoofprints, especially when hogs run or step into soft sand or mud. Trackways normally show an indirect register of the rear foot onto the front footprint in a diagonal walking pattern but can also display a full range of gaits, from a slow walk to full gallop. With repeated use of pathways, trackways become trails, although I am unsure if hogs are merely using and expanding the previously existing trails of white-tailed deer, if they are blazing their own, or if it is a combination of the two. (I suspect the last of these is the most likely.)

Rooting pits are broad but shallow depressions, as much as 5 meters (16 feet) wide and more than 30 centimeters (1 foot) deep. These pits are the direct result of feral hogs digging for food. In most instances I think they are looking for fungi and plant roots when they make these pits, but they probably also eat insect larvae, lizards, small mammals, and other animals that live in burrows. Pits are typically in maritime forests and back-dune meadows, but I have seen them in salt marshes and dunes and, most surprisingly, in the intertidal areas of beaches. What are they seeking and eating in beach sands? It could be anything dead and buried that might be giving off an odor. I have even seen their tracks directly associated with the shattered carapaces of horseshoe crabs, a hog-menu item that never would have occurred to me if I had not seen these traces. In salt marshes their pits hurt the chances of smooth cordgrass growth, leaving bare patches in what would normally be continuous expanses of this iconic and essential plant.[22]

Wallows are similar in size and appearance to rooting pits but have a different purpose, which is to provide hogs with relief from both Georgia summer heat and the biting insects that invariably go with it. These structures are often near freshwater wetlands in island interiors, but I have also observed them next to low salt marshes, on the salt flats of high marshes, and on storm-washover fans. If wallows intersect the local water table, they make attractive little ponds for mosquitoes to breed, meaning these hog traces also indirectly contribute to the potential spread of mosquito-borne diseases.[23]

Hog feces, consisting of Milk Dud–sized pellets, are understandably confused for those of white-tailed deer. Hog feces, though, tend to aggregate in clusters. Another way to distinguish a hog's scat from those of deer is through contents. Although most feral-hog feces I've seen have mostly vegetation, the extremely varied diets of their makers mean that you can expect nearly anything to show up in their scat: get ready to be surprised.

Which hog traces would make it into the fossil record? I bet at least some of their tracks would get preserved, based on their sheer abundance and ubiquity in nearly every sedimentary environment of a Georgia barrier island. Other likely traces amenable to fossilization would be their pits and wallows, although their broad size and shallow depths would render them difficult to recognize unless directly associated with hog tracks. Feces would be the least likely to make it into the fossil record as coprolites, unless these contained a fair amount of bone, eggshells, or other mineralized stuff.

What is there to do about these hogs, and how can we decrease the impacts of their traces? As most people know, pigs are wonderful, magical animals domesticated specifically for their versatile animal protein. Hence one solution is more active year-round hunting of hogs, using them to supplement breakfasts, lunches, and dinners of people living on the Georgia coast. This solution would comprise a neat blend of reducing a harmful feral species while encouraging a chic "locavore" label on such food. Year-round residents on Sapelo Island living in Hog Hammock were way ahead of such trends by hunting pigs on their island. Based on my estimates, the people there have done a fine job of managing the population, as Sapelo has the fewest tracks and other signs of hogs I have seen on any of the islands plagued by them.

But the sheer numbers of hogs on other islands, such as Cumberland, St. Catherines, and Ossabaw, would likely require a more systematic slaughter to make a dent in their numbers. This approach, however, would probably deter ecotourism and research unrelated to hog hunting. (For example, abundant firearms and bird watching can be an uneasy mix.) Another possible source of mitigation is the introduction of native predators, although this would likely not result in complete eradication. One instance of predator control is on Cumberland Island. There a population of bobcats (*Lynx rufus*) was introduced in the late 1980s with the main purpose of controlling the population of white-tailed deer.[24] These cats are not on venison-only diets, though, and also take a toll on juvenile feral hogs; still, enough piglets escape predation to reach adulthood and make more hogs. Another possible natural predator is the red wolf (*Canis rufus*), which could be reintroduced to islands with small human populations. These pack-hunting predators, which were native to the southeastern United States before their extirpation by fearful

FIGURE 39. The large-scale traces of feral hogs as ecosystem disruptors. *Left*, digging, rooting, feeding, and trampling of a back-dune meadow on St. Catherines Island. *Right*, pits showing the same behaviors, but on a remote beach on the north end of Cumberland Island.

Europeans and Americans, would likely reduce feral-hog populations too.[25] But just how much of an impact they would render is hard to predict. In the meantime alligators continue to be the main predators of hogs on Ossabaw, St. Catherines, Sapelo, and Cumberland Islands, snatching unwary victims that get too close to the water for a drink or swim. Unfortunately, alligators seemingly have only a minor effect on overall hog populations on Ossabaw and St. Catherines, meaning we cannot rely completely on our archosaurian friends for pest control.

For now we know that feral hogs on the Georgia barrier islands have significant effects on their ecology and geology. We also know they will continue having an impact until creative solutions are proposed and implemented to reduce and otherwise manage their numbers. In the meantime these invasive species present opportunities for us to study their traces, learn more about their unseen behaviors, and compare these behaviors with their evolutionary histories. Although I am not thrilled to know that some of my most beloved of the Georgia barrier islands are living experiments in the invasion ecology of feral hogs, I take small solace in the important ecological lessons we learn from them, ones that might not be available in the wide-open spaces of mainland Georgia and elsewhere.

22

Redbays and Ambrosia Beetles

Sometimes dullness stands out. In this instance the brown leaves in front of us contrasted with the more usual vibrant greens in the maritime forest of Cumberland Island, their difference inviting scrutiny. Curious—as I had never seen these plants in this state—I reached for and felt a few of the lance-like leaves alternating on branches in front of me. They were tough and brittle and broke easily. I cupped my hands to hold the pieces, brought them up to my nose, and inhaled. The smell was mustier than expected, lacking the spice I normally associate with this plant. When I looked farther into the forest, I saw more browned examples mixed with the living, lending to a foreboding sense of doom for those who had not yet been caught by whatever killed the others.

These departed plant parts belonged to redbay (*Persea borbonia*), a common species of understory plant in maritime forests of the Georgia coast. My observations of these dead and dying plants happened in March 2013, while leading a group of Emory students on a spring-break field course, in which we visited six islands (Cumberland, Jekyll, St. Simons, Little St. Simons, Sapelo, and St. Catherines) in nine days. Because our itinerary had us moving from south to north, Cumberland was our first island, but we were there just for the day, having arrived on the ferry in the morning and riding back to the mainland by late afternoon. Given such limited time, I wanted to maximize the quality of their educational experience. So rather than subjecting my students to impassioned soliloquys bespeaking of, say, coquina clams, I arranged beforehand for ace naturalist Carol Ruckdeschel to talk with the students about the natural history of Cumberland. Thankfully, she graciously accepted my request, and I was delighted to see her when she met us at the dock that morning.

Following a brief introduction by me of Carol to the students, we gathered around a few picnic tables near the dock, notebooks and writing implements emerged, and she gave my students a marvelous synopsis of the Cumberland Island ecosystems. In this brief lecture Carol highlighted the tensions between previous human impacts on those ecosystems and how those actions can collide with the now. Only ten minutes into her talking, an errant feral horse emerged from the forest behind her and lumbered past our busy note-taking group, seemingly making her point. Following this interruption the lecture became more mobile and interactive when we went for a walk on a trail through the maritime forest, then traveled on a boardwalk over back-dune meadows. Once we reached the ecotone between the back-dune meadows and the coastal dunes, Carol bid us bon voyage, and my students were stuck with just me for the rest of the day.

It was during our time in the forest when Carol pointed out the dead redbays and explained what was then known about their demise. An exotic wood-boring beetle, the ambrosia beetle, which is now nicknamed the "redbay ambrosia beetle," arrived on the island in the previous decade.[1] Its exact advent on Cumberland is unknown, and because the beetle was so small (only 2–3 millimeters long), its initial effects may have gone unnoticed until a significant number of redbays died there in 2006.[2] We do know, however, that it was detected on St. Catherines Island as early as 2004 and had spread to all other Georgia barrier islands by 2008.[3] Hence in 2013 we were seeing results of only the first few years of its invasion on Cumberland and other islands.

Since then I have tried to learn more about both redbay and its alien attacker. For instance, I always thought of redbay as a bush or shrub, but foresters actually consider it a tree. This upgrading is because redbay can grow to almost 20 meters (66 feet) tall, which means its topmost leaves could be picked from the window of a five-story building. On the islands and parts of the lower coastal plain, though, competing plants like live oaks and saw palmettos limit its height. Take a closer look at a redbay, and you will see its smooth-margined leaves are lance-like but slightly oval, show a darker green on top surfaces, and alternate along branches, varying from 8–15 centimeters (3–6 inches) long. Its bark is red-brown and ridged. Although redbay is an evergreen, it is also a flowering plant, producing both flowers and fruits. Its flowers are small (3–6 millimeters wide) and pale yellow, and its deep-blue to black fruit is not much bigger, measuring only about 1–2 centimeters

(0.5–0.8 inches) long. Mostly birds eat this fruit, which is purportedly quite bitter to us. Native Americans used redbay leaves and roots for a variety of medicinal purposes, whereas Europeans used its wood for boats, cabinets, and other woodwork.[4] Also, if you have ever eaten a traditional gumbo or low-country boil, entire redbay leaves were likely used to season it.

At some point in my learning about redbay and its decline, I was surprised to learn this seemingly simple scenario of native victim and imported villain was far more complex, requiring a look below the surface—that is, the surface of the redbays themselves. Redbay ambrosia beetles, like all beetles in its evolutionarily linked group (Scolytinae), are bark beetles, which make their living by boring beneath the bark covering shrubs and trees.[5] Unlike moon snails, bark beetles do not carry personal drills with them but instead chew through wood with their mandibles. Moreover, the adults are not actually eating the wood for food but are instead constructing living spaces, nurseries, and gardens. The living spaces are tunnels and galleries for the adults, the nurseries are hollowed-out spaces for female beetles to lay eggs, and the gardens are for growing a fungus that serves as food for hatched larvae and adults.[6] These fungal associates are often tagged to each beetle species, and the redbay ambrosia beetles' cohort is laurel wilt.[7] Bark beetles are rare among insects in growing their own fungal food; only some ants and termites have likewise taken up gardening as a means of sustenance.

To make matters more complicated, not all bark beetles use this form of symbiosis, but the ones that do have the greatest effect, an ecological tag-team effort that can devastate entire forests. Because adult bark beetles can fly and fungal spores are their carry-ons, both native and imported species spread quickly. As a result, bark beetles are responsible for killing vast swaths of forests composed of conifers, such as pines, spruce, hemlock, and firs.[8] Deciduous trees (like redbay) are not exempt from beetle-fungi infestations either. Another example of such symbiotic illnesses is Dutch elm disease, caused by at least three species of bark beetles and various species of *Ophiostoma* fungi, which first affected elms in Europe before spreading to North America.[9]

Ichnologically speaking, bark-beetle borings look like lovely works of art, intricate carvings that follow surfaces just underneath the outermost bark. Despite knowing what some wood-boring beetles can do to trees and eventually entire forests, I cannot help but admire these mandibular creations. Hatching galleries are especially marvelous, bearing a single straight tunnel in their centers, from which radiate dozens of meandering, smaller-diameter tunnels, looking like spokes on an oblong wheel.[10] Follow these tunnels along their lengths, though, and you will also notice a slight widening toward their ends. This expansion is because they represent emergence tunnels of newly hatched larvae, reflecting bodily growth of larvae on their journey into the wood and on the way to pupation and adulthood.

Consequences of this bioerosion by larvae and adult beetles are other subtler traces. For instance, one way to know ambrosia beetles have invaded a redbay without cutting it open is to look for threadlike clusters of chewed wood on its outer bark. This is frass, which generally refers to processed wood that either has or has not passed through an insect's gut.[11] Remember, for ambrosia and many other bark beetles, wood is not food, but fungi are, meaning their frass is not fecal, but sawdust.

Because of their distinctive patterns, bark beetles both past and present are detectable by their traces. I have marveled at fossil examples of beetle borings in petrified wood from the Late Triassic period (about two hundred million years ago) and often contemplated how these beetles were busily eroding dead and dying trees as oblivious dinosaurs walked by.[12] Still, not all dinosaurs ignored these beetles or other wood-boring insects. In a 2007 study paleontologist Karen Chin documented how wood fragments in dinosaur coprolites of Montana from seventy five million years ago showed signs of having been weakened by fungi.[13] This meant the wood consumed by these dinosaurs was probably already dead. Why would a dinosaur eat dead wood? This is a good question, as this stuff is mostly devoid of usable nutrition. Hence Chin speculated that fungi and insects in the wood might have enhanced it, providing a boost of protein and other nutrients that made it worth eating. Much later this practice evolved in modern wood-drilling and insect-consuming dinosaurs, which some people stubbornly refer to as "woodpeckers."

Wood-eating behavior in insects and other arthropods goes back to nearly twice as long as dinosaurs in geologic time and to the very beginning of woody tissues in plants, which we know thanks to trace fossils. For example, fossilized frass and tunnels in leaves and woody tissues from Devonian period rocks (about four hundred million years ago) show that arthropods—perhaps mites, perhaps early insects—had already begun using plants as food sources and homes.[14] Cockroaches, which evolved in the Carboniferous period about three hundred million years ago, even gave rise to the most

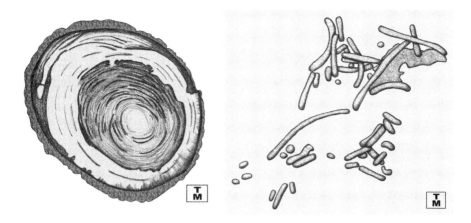

FIGURE 40. Invasive symbionts of redbay ambrosia beetles and laurel wilt, coconspirators in the demise of redbay. *Left*, blackened rings in a cross-section of redbay, a sure sign of laurel-wilt infection, brought in by wood-drilling ambrosia beetles. *Right*, ambrosia-beetle frass (chewed wood) pushed out of tunnels, helping their gardens of laurel wilt grow. Illustrations by Anthony J. Martin based on photos by Michael Flores, University of Florida.

famous of wood-boring animals, termites.[15] For wood-boring beetles, bark chewing has been a fact of their lives since at least the Triassic period, and they were joined later in the Cretaceous period by bees.[16]

Evidence for fungal infections of plants in the fossil record is less common, but remarkably modern-looking fungi body fossils are in Silurian-Devonian rocks from more than four hundred million years ago.[17] These fossils hint at the early symbiosis between fungi and land plants, the latter of which were still colonizing formerly barren rocky landscapes and looking nothing like the verdant prairies, forests, or jungles of today. Although most people (including me) commonly refer to mushrooms as "veggies," these and other fungi are more closely related to animals than plants.[18] Fungi neatly demonstrate this relationship by not photosynthesizing but instead consuming both plants and animals.

Given this deep-time perspective, one might argue that the whole-scale destruction of redbay on the Georgia coast by a beetle and a fungus is insignificant and perfectly natural. After all, modern fungi and wood-boring insects are among the most important nutrient recyclers of forests, helping to return the material essence of dying and dead trees to the forest to future generations of their communities. Also, nearly all fungi, plants, or animals that have lived are extinct, and five mass extinctions happened in the geologic past. So what's the big deal with losing this one species on the Georgia coast, especially when polar bears, pandas, rhinos, and other charismatic megafauna need saving first?

For one, an impending extinction for redbay on the coast is not a part of a natural extinction pattern and would not have been possible without human help. Ambrosia beetles and laurel wilt evolved together in Asia, whereas redbay evolved in the southeastern part of North America.[19] There all of redbay's defenses naturally selected for insects and fungi of that region over at least the past few millions of years. Hence when people brought wood from Asian ecosystems with breeding populations of ambrosia beetles and their fungal cohorts through Georgia ports, they upset the ecological détente between redbay and its enemies. It was like importing an overeager recycling crew that became far too proactive by emptying bottles and cans of their contents before they could be used. Redbay was adapted well enough against its normal enemies that it stayed alive and thrived as an integral part of maritime forests on the Georgia barrier islands and the nearby lower coastal plain. But now it may be reaching a point of no return, with diminished numbers and rarity as its most optimistic scenario.[20] A maritime forest without redbay would be a less complete version of what once was and quite likely would be a less functional one.

Granted, these are the ecological and aesthetic aspects, which economic fundamentalists might readily dismiss with the wave of a hand, telling us this is a small price to pay in the scheme of global commerce. So this is when you may recall my mentioning the use of redbay as a key ingredient in traditional Southern cuisines. Yet if this blander future is not enough to cause pause, then think of a world without guacamole. Amazingly, one of the closest relatives to redbay with its bitter fruits is *Persea americana*, the avocado. Indeed, at the time I write this, laurel wilt and ambrosia beetles were already assaulting avocado trees in north Florida and projected to spread to south Florida in the near future.[21] With no effective means of combating the spread of these partners in ecological crime, domestic supplies may dwindle and avocados may accordingly become much more expensive, perhaps affordable only to eccentric tech billionaires. This is yet another example of how ichnology, ecology, and invasive species intertwine, even if the perpetrators are teeny beetles and musty fungi rather than hogs, horses, and cattle.

When all is said and done, though, redbay ambrosia beetles and laurel wilt are just two small-bit players in a much grander drama being staged on the Georgia coast, a production with a cast of billions and no clear ending. This realization hit home for me in late February of 2018—almost exactly five years after taking my students to Cumberland Island—as I led another group of Emory students on a Sapelo Island field trip. Within just five minutes of walking a trail I had traversed with generations of student bodies for nearly twenty years, this group for the first time saw me stop and stare at a clearing that did not belong there. A pine forest I had admired for nearly a third of my life was no more, represented only by trunks and limbs of former trees strewn between stumps and barren, needleless poles.

As I struggled to comprehend this anomalous sight, I wondered aloud whether this former forest was a result of Hurricane Irma, which had struck the Georgia coast less than six months before. Then I saw a sign—that is, it was posted literally as a sign next to the trail asking exactly the question on my mind: "What Happened Here?" The answer was that the pine trees had been attacked by southern pine beetles (*Dendroctonus frontalis*) and were cut to keep the beetles from dispersing to other forests on the island: kind of like destroying a village to save a city. As one might surmise from their name, southern pine beetles are native to the southeastern United States, and, like other bark beetles, they bore extensive tunnel systems underneath the outer bark of native pines.[22] These beetles similarly have a native fungal cohort, the blue-stain fungi (*Grosmannia clavigera*), which infect pine tissues and hasten tree deaths. Although I had read reports about pine forests on the mainland overwhelmed by this conjoined threat, to see its ill effects on my most beloved of Georgia barrier islands made it all too depressingly real.

Regardless of whether redbays and pines of the maritime forests are felled by enemies foreign or domestic (respectively), their foes have a common ally, one with effects more readily apparent away from those forests and along the shore. This enabler is climate change, which has accelerated the reproductive cycles of these and other bark beetles so that the trees may not be able to adapt quickly enough.[23] Would this be the first time that humans had radically altered maritime forests? Certainly not. But the original human inhabitants of the islands modified these ecosystems in more sustainable ways, such as the use of redbay and other forest plants for improving their lives, which allowed for the development of traditional ecological sciences.

So now let us look toward the not-so-distant past—only about five thousand years after the last major shift in climate—to see what significant traces the first peoples of the Georgia barrier islands left behind. What do these clues tell us about this ecologically interloping species adapting natural resources to meet its daily needs? Through such traces we may then learn how the past may apply to solving the problems of the Georgia coast and its uncertain future.

23

Shell Rings and Tabby Ruins

First the tabby ruins, then the shell ring.

However nonsensical such a phrase may sound to the uninitiated, for me it summarizes the agenda of a typical Sunday-afternoon field trip on Sapelo Island. Whenever I've taken my Emory students to Sapelo for a too-short weekend visit there, we use all of Saturday to experience a nature trail on the south end of the island. This trail gives students a taste of maritime forests, salt marshes, tidal creeks, back-dune meadows, dunes, and a beach, all captured in a few-mile trek. On Sunday mornings we head north to see Cabretta Beach and its relict marsh and perhaps the Pleistocene outcrop at Raccoon Bluff in the northeast corner of the island before cutting west to the northwestern corner of the island.

Along this edge of Sapelo is a grassy clearing just off the sandy road, backed by salt marshes alongside the Duplin River tidal channel. Toward the back of the open area is a 1930s Sears and Roebuck mail-order house, which sits next to a restored barn with walls of cemented oyster shells. These two intact structures, however, are not the main reason for our visit. We instead spend time walking around and between ruins, their collapsed walls marking the sites of roofless buildings. Most buildings were used to contain enslaved people, whose labor and lives drove agriculture on this part of Sapelo and other Georgia barrier islands in the late eighteenth and early nineteenth century.[1] At some point in the postslavery past, locals nicknamed this location "Chocolate Plantation." This place is called this not because of plants bearing ready-to-pick candies but instead because of Chucalate, the name of a Guale (Native American) village that was once there.[2] My students always remember its name because of how it inspires confectionary visions of

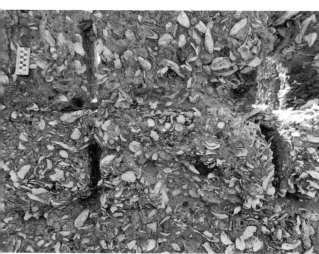

FIGURE 41. Tabby ruins on the site of a former plantation on Sapelo Island at the misleadingly named "Chocolate Plantation." *Top*, perspective of a wall from one of the buildings, expressing its shelly framework. *Left*, close-up of a wall composed of bivalve shells—mostly eastern oysters—held together by lime cement.

Wonka-like tours, but such whimsical thoughts are quickly replaced by the sobering reality of these grounds, the buildings, and their horrid history.

The barn and remnants of other buildings at Chocolate Plantation still stand mostly because of oysters. The shells of these molluscans serve as minute "bricks," interlocking with one another but further stuck together by cement also derived from shells, a building method called "tabby." This odd name—more associated with cats than houses—may have come from Spanish, with *tapia* meaning "mud wall," but it also may be from an Arabic word (*tabi*), perhaps placing its origins in northwest Africa.[3] Tabby cement is composed of oyster shells fired at high temperatures, ground into a powder (lime), and mixed with water, quartz sand, and ash before the addition of entire and broken shells. Modern styles of cement were not widely adapted until the twentieth century, and the Georgia barrier islands were isolated enough that European building supplies were not easily shipped there, so their colonizers used local materials.[4] These people also threw in other binding and framework substances, which varied according to what they had on hand. For instance, I have sent my students on "treasure hunts" to look in the walls for bits of charcoal, brick, ceramics, glass, or hair from domestic animals. During one visit a colleague of mine even found part of a smoking pipe, which through its internment managed to stay tobacco-free for more than two hundred years.

While there I also look closely at the shells for borings, such as the tiny, bobby-pin-sized holes left by sponges or bryozoans. The presence of such traces are indirect indicators of how long the oyster shells were exposed in marine-influenced waters to allow sponges and bryozoans to settle on them. But most shells lack these borings, implying they likely spent most of their time after death out of the water, either buried or exposed to air. If I notice this, I share it with my students as a little clue for their pondering later.

Following this short visit to Chocolate Plantation, we resume travel on a sandy road north of it to encounter a bigger mystery. On the way there the maritime forest on either side of the road feels a bit wilder and farther removed from modern places and times, as if we are traveling back in time. My students, sitting on hard benches in the back of an open-bed truck, cry, "Duck!" more often as we encounter increasing numbers of low-lying muscadine vines and live-oak branches draped across the road. No signs or other recent human markings tell me where to pause. It becomes more of a feeling that we have arrived, which I test and verify by looking left for a subtle rise in the landscape underneath the thick forest.

There. We stop, and the abrupt silence of the stilled truck helps set a somber mood. I ask everyone to walk on a human-sized path across this land and behind me and experience it with a minimum of talking, observing their surroundings before reflecting and sharing what we experienced. In leading them I also try to cultivate a sense of reverence and respect, which is only appropriate, as we are walking atop a place where many people lived and died.

What we all notice while walking on the path is how it traverses a slope that gradually rises to a peak of about 3 meters (10 feet). For those people who look to their right and left, they may discern how this high spot is part of a ridge that curves away on either side and into the forest. But I completely understand if they do not catch this continuity, as mature live oaks, saw palmettos, redbay, and many other plants hide its surface. From this high we descend and travel across a low, flat area for about 70 meters (230 feet) before encountering a dramatic cut through the land aligned with the trail, affording an open-door view of a salt marsh just beyond and below. As we walk through the cut, we see it is about 3 meters tall on either side, that it is inside the same curving ridge observed before, and that the ridge is filled with shells. Unlike the tabby, only soil and compaction holds this structure together. As a result, some shells have tumbled out of the sides, inviting us to pick them up and examine them. Most are oysters, although we occasionally pick up a southern quahog (*Mercenaria mercenaria*), whelk, or other molluscan species in the mix. After less than a minute, all the students soon realize they are inside one tiny slice of an enormous ring composed of millions of shells, and they are peering into its interior, like standing where a sliver of a piece was taken from an enormous Bundt cake.

With each visit I purposefully do not tell them about this shell ring beforehand, which I've found better enables a sense of discovery and sparks their curiosities and imaginations. Following their recognition of the shell ring as a huge human trace, I then ask questions, such as, How large is it? How old is it? Who made it? Why is it composed mostly of oysters? How did the oysters get there? How long did it take to form? And so on, until I check their answers against what is known about the ring from scientific investigations there since the 1920s.

People made the Sapelo shell ring three thousand to four thousand years ago, predating the Spanish arrival to the Georgia coast by an order of magnitude and testifying to a long human occupation of the Georgia

FIGURE 42. An inside look at the western edge of the largest and best-expressed shell ring in the Sapelo Island shell-ring complex, with mostly eastern oysters represented. Exposure is from an archaeological investigation, not from mining for tabby buildings, which is another human trace on the south side of the ring. Spousal unit (Ruth Schowalter) for scale.

coast.[5] At about 100 meters (330 feet) wide and 3 meters (10 feet) tall, its size also tells of both cooperation and time. This was not the accomplishment of one rugged individualist shucking a dozen oysters daily for a lifetime but of collective action by a community working together over many generations. After some discussion, we all agree that the oysters and other items composing the ring (including deer and fish bones, which I mention later) probably represent food scraps from many meals. Such deposits on the Georgia barrier islands and elsewhere are called middens, and when they are composed mostly of molluscan shells, we more specifically call them shell middens.[6]

Given that many people lived on the Georgia coast for thousands of years, it is then not surprising to know the Sapelo shell ring is neither unique nor alone, and similar rings on other Georgia barrier islands were actually quite common. Indeed, two smaller rings were near the Sapelo one, meaning all three actually constitute a ring complex.[7] Unfortunately, the two other rings are no longer easily discernable at the surface, although modern technological tools, such as LiDAR (Light Detection and Ranging, using laser scans) and ground-penetrating radar, helped in mapping their locations. Similar shell rings are documented elsewhere from coastal Georgia, South Carolina, and both Atlantic and Gulf coasts of Florida.[8]

When Spanish ships first sailed into Georgia coast waterways in the early sixteenth century, the Guale and Mocama peoples were living on Sapelo and other islands just south and north of it.[9] Farther south were the Timucuan people, who lived on Cumberland and coastal environments in a territory the Spanish later called Florida.[10] But we do not know for sure if these Native American inhabitants actually descended from the shell-ring builders or whether their ancestors arrived after the ring makers had already come and gone.

Europeans and their descendants in Georgia recognized shell rings on the Georgia barrier islands in the eighteenth century, and more were being rediscovered up into the twenty-first century. Rings are so far known from Ossabaw, St. Catherines, Sapelo, and St. Simons Islands, although smaller middens are apparently on all the islands.[11] Not all were true "rings," either, as some were more like horseshoes, with one end open. Nonetheless, the circular plan of the Sapelo shell ring provokes a basic question during discussions with my students: why a ring? For instance, a horseshoe design actually makes more sense if people were inclined to move in and out of its interior. So why not make more *U*'s instead of *O*'s? Granted, a three-meter climb or descent was not such a big deal for people who did not have to drive to a gym for their exercise. But ease of access certainly would have been improved by keeping one side open.

Here is where reasonable hypotheses (testable and falsifiable) collide with speculative fiction (untestable and fanciful), as we struggle with concepts of form and function, while also taking into account more abstract and less predictable facets of human thought, such as spirituality. (Our latter trait is why I try not to make behavioral analogies between humans and, say, lobsters.) Anthropologists often joke that if they are stymied about the purpose of a particular human-made object, then the default explanation is that it must be religious. Sadly, we cannot ask the Guale, Timucuan, or nearby native peoples directly either, as they left no written records of what happened there before Spanish ships dropped anchors in their waters. Oral traditions, which also could have at least preserved stories, were lost as representatives of native cultures were killed, enslaved, sickened from European diseases, or assimilated into imported cultures. We do, however, still have traces of Guale thoughts and creativity from intricate designs they stamped on fire-tempered ceramic pottery, the remains of which are the oldest known in North America.[12] Alas, none of these traces tell us the "why" or "how" of shell rings either.

In the case of the Sapelo shell ring and others like it, these structures might actually be a combination of middens from feeding and ceremonial or monumental structures built around circular plots of dwellings and villages.[13] This sort of city planning also would have made ring construction much easier than if attempted from outside a village. People could simply walk out their back doors (so to speak) after yet another roasted-oyster dinner and dump the shells onto a pile behind their homes. Recent archaeological studies of the two shell rings on St. Catherines Island also show how shells were deposited in definite layers.[14] This layering implies that people likely devoted more effort to building them during certain times of the year, which in turn could reflect seasonal changes in food supplies, rituals, or both. Moreover, chemical signatures from shells in the main Sapelo ring indicate they came from waters with widely varying salinities, rather than just one or two nearby places.[15] That means people did not walk out to nearby marshes for their food (which would have been unsustainable) but traveled in boats all around the island to gather it. Given the great tidal exchanges along the Georgia coast, grocery runs also probably had to be timed carefully in accordance with those daily cycles.

Did the rings serve any purposes other than as refuse piles or visible signs of gratitude for abundant food resources provided by the local ecosystems? Other ideas my students provide are more notional and often boil down to "humans versus humans" or "humans versus nature" scenarios. In the first we think of how a three-meter-tall barrier around a village might have served as a fortification, deterring raids or other assaults by people of neighboring villages. Of course, human history shows repeatedly how walls—big, small, or metaphorical—eventually fail. Also, peaceful harmony and cooperation between different settlements might have been more common in Guale culture than we can imagine, especially if our perceptions are being influenced by the dominant socioeconomic systems of today.

Yet another consideration, and a practical one, is the importance of getting a good view on a flat island, one that does not require climbing a tree. After all, this and other shell rings may have been taller in the past, perhaps reaching stratospheric heights of seven to eight meters (twenty-three to twenty-six feet). Spotting people from afar, whether friends, foes, friendly foes, or potential bad dates, would have provided additional advantages to whatever communities lived inside a ring.

When thinking about threats from nature, though, we first think about wild animals—no, not marauding congresses of alligators but rather mammalian concerns, such as black bears (*Ursus americanus*), cougars (*Puma concolor*), and wolves (*Canis lupus*). All three of these carnivores lived on either the mainland or islands during Guale occupation. But the advent of Europeans and Americans led to the extirpation of wolves in the southeastern part of North America, and they nearly eradicated cougars too.[16] Black bears hung on; indeed, rare individuals were reported on Georgia barrier islands as recently as the 1930s.[17] Realistically, though, even the largest of the Georgia barrier islands could not have supported significant populations of bears or other large carnivorous mammals.

Modern people insulated from predation also like to imagine themselves as special treats for ravenous carnivores that seemingly can't wait to get some of that sweet, delicious human flesh. Yet the reality is that we are on the last page of their menus, in fine print, and ordered only when ecosystem kitchens have run out of every other dish. When one considers how abundant supplies of deer, bison, and even eastern elk (*Cervus canadensis*

canadensis) were on the mainland until the sixteenth century, why would large carnivores need to swim to an island for its food and then go after hostile bipeds that had mastered fire and wielded pointy sticks?[18] Bears, on the other hand (or paw), with their scent-motivated ways and voracious omnivory, still might have been attracted to human settlements as places to stock up on a wide variety of prepared foodstuffs. So, yes, I could certainly see bear deterrence as one of many possible functions of a tall, ring-shaped enclosure. Yet it certainly would not have been high on any list of reasons for such a massive commitment.

Given enough questioning and discussion, though, my students eventually think of a much more likely nature danger requiring the construction of a high barrier, and it is meteorological. For instance, think of hurricanes, nor'easters, and other storms that tend to drop or transport large volumes of water in short periods. I prompt this possibility by pointing to the geographic position of the Sapelo shell-ring complex on the northwesternmost corner of the island. This location means it would have been the least likely to encounter waves from tropical storms making landfall from the southeast. Interestingly, native peoples on the Georgia barrier islands were not into beachfront homes and kept themselves relatively far from the sandy shorelines of the barrier islands. So, unlike current coastal developments, they neatly avoided any potential problems suddenly imposed by five-meter-tall storm surges.

Because the Sapelo shell ring is so prominent to its visitors, one might think that simply elevating the ground in round patterns was the most visible way that native peoples changed their landscapes. Yet even buried shell rings and other middens also altered terrestrial ecosystems, leaving now-subtle traces of a former human presence. One of the easiest ways to understand how these pre-European-conquest people who "lived off the land" modified their environments is to ask yourself how often you normally encounter thick deposits of oysters in a maritime forest. Humans transporting calcium-carbonate shells from salt marshes to inland environments over many generations and leaving them buried for thousands of years radically changed island soils. As most gardeners know, lime, which is powdered calcium carbonate (derived appropriately enough from limestone), is used to buffer soils by bringing down their acidity. Molluscan shells in acidic soils regularly exposed to slightly acidic rain and groundwater also dissolve. Think of it like a soil taking massive quantities of antacids, in which shell deposits increase soil pH by liberating calcium and bicarbonate and even later precipitating calcium carbonate. I once saw a radical example of the latter phenomenon on Cumberland Island, where a shell midden exposed on a coastal bank contributed to the development of calcite-coated tree roots below it. Once modified, calcium- and bicarbonate-rich soils also become more basic (and less acidic). These soils in turn select for calcium-loving plants better adapted to such conditions, allowing them to sprout, grow, and thrive.[19]

Amazingly, an understanding of the botanical effects of shell rings led to the discovery of a previously unnoticed ring on St. Catherines Island in the early twenty-first century. One day in 2008 archaeologist David Hurst Thomas and the St. Catherines island manager, Royce Hayes, were prospecting for possible archaeological sites in a St. Catherines maritime forest in the southeastern part of the island. At some point in their wanderings, Royce noticed a concentration of plants he knew preferred calcium-rich soils. Even better, the plants followed a circular pattern, a coincidence that was not one. Later that year David led a crew in digging a pit in part of the pattern, and they found plenty of oyster shells and other remains of former food items below. The ring, which has since been mapped and intensively studied, was dubbed the McQueen shell ring for its closeness to the McQueen Inlet on that side of the island.[20] It was as large as one previously discovered on the northwest side of the island, bearing a 10-meter (33-feet) wide wall and a 30–40 meter (100–130 feet) wide interior.

How did such a massive human-made structure on St. Catherines escape our notice until now? At least one reason is because it and other less-obvious shell rings survived deliberate destruction. Following inspired discussions with my students at the cut through the Sapelo shell ring, I often lead them on another path up onto its top, from where we curve to the southeast. We wind around live oaks and thick underbrush on the rampart until I stop everyone at the southern edge of the ring and point to a 20-meter (70-foot) wide depression below that disrupts the integrity of the ring. After a few leading questions, we arrive at an answer, one that we dislike intensely, because it is true.

Remember the tabby ruins? Where do you think agriculturally inclined Europeans and Americans found so many shells to use for making their tabby buildings? Did you think they traveled to tidal creeks and marshes all around Sapelo, collected oyster shells, and brought them back to the plantation, trip after trip? No, they quarried them, from a nearby multi-generational mother lode unwittingly supplied by extinct people from three thousand to four thousand years in the past. The depression on the southern margin of the shell ring is a pit made by surface mining, marking where enslaved people of African descent about two hundred years ago dug into it, extracted its shells, and moved them a few kilometers south to the plantation site. Before then, in the 1730s, British general James Edward Oglethorpe directed people to do the same on St. Simons Island.[21] There shell middens were obliterated to provide raw material for building the military outpost of Fort Frederica, which had a facsimile of an English village in its center. Other shell deposits on the north end of St. Simons Island at Cannon Point were exploited for a plantation there, with tabby ruins still standing in mute testimony to this plundering.[22]

In a sense, then, the tabby ruins of colonial times reflect the erasure of a once-thriving native presence and of how the people of the recent past—through their mining the remains of a deeper human past—left traces lasting centuries into their (and our) future. Given these connections between the traces of different times and peoples, I yearn for my students and others to better understand how our living with the Georgia coast also means living with intertwined ecological and human stories of what was and what will be, before and after our time.

24

Ballast of the Past

While strolling through the charming and historical city of Savannah, Georgia, in December 2015, I made sure to pay attention to the thousands of time machines below my feet. Yes, I know, everyone other than geologists stubbornly refer to such objects as "rocks." Fortunately, earth scientists don't have to limit their imaginations with such simplistic labels. These pieces of a prehuman past all have stories to tell of their origin, and sometimes their journeys even relate to our treatment of one another as human beings, bestowing lessons in morality and justice.

First of all, the rocks of Savannah do not really belong there. This is especially true for the cobbles on the north end of town dotting roads and walkways and reinforcing walls near the Savannah River. A quick glance at these stones by the geologically informed reveals how they are all foreign to this part of Georgia. Indeed, most are from across the Atlantic Ocean, and the majority probably originated in the British Isles. Yet they also have been part of Savannah history for at least a few hundred years. What are they, how did they get there, and why are they there?

These are ballast stones, rocks that filled the cargo holds of ships during the eighteenth and nineteenth centuries as they sailed across the Atlantic Ocean from England. Were these ships exporting rocks to eager colonists in the Americas who wished to collect nostalgic (and solid) reminders of their former homelands? No, ballast stones were used to keep ships weighted down, which helped stabilize them as they moved across seas both calm and rough.[1] Once a ship reached Savannah—which began as a British settlement in 1733—its crew would dump its geologic load and replace its relatively uneconomic value with agricultural goods grown in Georgia, such as rice, cotton, and indigo.[2] Those commodities then went across the ocean, where they

FIGURE 43. A cobble-lined street in the historical district of Savannah, built of rocks from another land and by people from another land, some of whom did not have a choice in building them.

were used for food or textiles. Meanwhile, the ballast stones were repur-posed as durable materials for streets, walls, and houses along the Savannah River, as well as in some of the older homes in the more upland historical district of Savannah.

Farther south, near the former port city of Darien, ships discarded bal-last stones alongside the banks of tidal channels, forming ballast islands.[3] Examples of these islands are visible on a ferry ride to and from Sapelo Island, and I often point them out to my field-tripping students during the voyage through Doboy Sound. The islands stand out as a string of forested hammocks on the eastern edge of what has been dubbed "Rock Island," eco-logical anomalies in an otherwise continuous expanse of marsh and open water. Although I have looked at satellite images of these islands, I have never stepped foot on any of them, which makes them feel remote and con-ceptually abstract to me.

In contrast, the rocks on the streets and in the walls of Savannah are easy to visit and study at leisure. They are remarkably varied, reflecting the geologic diversity of the United Kingdom and perhaps elsewhere in Europe. Igneous, metamorphic, and sedimentary rocks are all represented, but per-haps the most common is basalt. Basalt is a black, fine-grained extrusive igneous rock, formed by hot magma that cooled at or near the earth's sur-face, disgorged by a long-gone rift or volcano. Some of the basalt cobbles in Savannah are filled with holes, or vesicles; these are a result of gases bub-bling out of the original magma and lending to a more specific name for the rocks: vesicular basalts. But I also saw intrusive igneous rocks, their large, interlocking crystals telling of magmas cooled at great depth. At least one rock, though, was in turn intruded by basalt, defined as a clean, black band cutting across the older rock. This simple geologic relationship of one phenomenon (the basalt) cutting across another (the large-crystal intrusive rock) is appropriately dubbed "cross-cutting relations," neatly denoting rel-ative time: the intrusive rock existed first, then the extrusive rock.

Sedimentary rocks include sandstones, some of which were placed paral-lel to their original bedding, fitting like bricks in some of the walls above the streets. These sandstones bear properties superficially similar to sands of the Georgia barrier islands, composed of durable quartz sand sorted into an-gled layers called cross-bedding. At least a few other sedimentary rocks con-tained fossils, such as a limestone with gorgeous lengthwise sections and cross-sections of crinoid stems. Crinoids, which are echinoderms and dis-tantly related to sea stars, sea cucumbers, and sand dollars, are still around, although far less common than in the geologic past.[4] From these fossils I reckoned this rock was probably from the Carboniferous period toward the end of the Paleozoic era, representing a former sea from more than 300 million years ago. Another limestone near it contained what looked to me like oncolites, spherical to oval concentric structures made by the growth of cyanobacteria or algae around a small stone or fossil fragment.[5] Rocks bearing oncolites were common earlier in the Paleozoic era, say, 450–500 million years ago, meaning that this limestone was possibly older than the crinoid-bearing one.

Like most normal people, you are probably wondering how these ballast stones relate to ichnology. For instance, do any of the sedimentary rocks contain trace fossils? Perhaps they do, although I didn't see many convincing examples. Only one of the many rocks I examined had possible vertical burrows exposed as holes in a sandstone cobble. These holes likely represent cross-sections of burrows that continue down into the rocks; moreover, some holes were apparently paired, suggesting at least a few were originally U-shaped burrows. U-shaped burrows are easier to identify in lengthwise sections, which may show how the two tubes join each other in a rounded bottom.[6] Without this view, though, another clue to identifying a U-shaped burrow from its top or bottom is a dumbbell-shaped depression. With this shape the line (the bar) between the two holes (the weights) represents where sand collapsed between the two burrow limbs. Suspension-feeding animals, such as polychaete worms or small crustaceans, may have occupied such burrows hundreds of millions of years ago.[7] While looking at these, I thought of similar modern examples that might be in offshore sands of the Georgia coast, a short distance from where I stood in Savannah.

Yet another trace was there, one much larger and more conceptual than what can be discovered in a single rock. Think of how these thousands of ballast stones collectively represent a human trace, tangible evidence of a grand transference of geologic heritage from one continent to another. From more of a moral perspective, however, these ballast stones are also a trace of slavery. The labor of enslaved people—abducted from their homes in western Africa and, like ballast stones, packed into cargo holds on ships and taken to a foreign land—produced the agricultural goods that went back in ships to Europe.

Although slavery was at first banned in Savannah after the city's founding, it was allowed soon afterward, starting in 1750. It continued after U.S. independence from the British and persisted through the latter part of the

FIGURE 44. Fossils from afar in both time and space among the ballast stones of Savannah, which now line some of the city's streets and walls. *Left*, limestone bearing abundant skeletal debris of crinoids, animals that lived in a shallow sea more than three hundred million years ago. *Right*, sandstone hosting numerous holes that are likely cross-sections of burrows made by wormlike animals, also in a shallow sea. Notice how some holes are paired and joined, suggesting these are the tops of U-shaped burrows.

eighteenth century and more than half of the nineteenth century.[8] During that time Savannah was among the most productive ports in the world for the shipping of rice, cotton, and timber.[9] The heinous exploitation of human lives during antebellum times continued until the advent of the U.S. Civil War in the 1860s. This meant many ships arrived over the years, bringing their ballast stones and taking back cotton, rice, and other products of this cruel labor. Meanwhile, enslaved people were also used to construct many of the streets, walls, and homes in Savannah composed of ballast stones and bricks. Some of the red bricks in these old city walls even bear the finger impressions of captive people who shaped their once-soft clay.

Thus there would be far fewer ballast stones on the streets and in the walls of Savannah, and no ballast islands near Sapelo Island, if not for this part of British and U.S. history. The legacy of these stones also links to the family lineages of millions of African Americans who survived their enforced journeys, whether these people live in Savannah, other parts of Georgia, throughout the United States, or abroad. As I tread these rocks in the streets of Savannah, I am mindful of how their physical weight is also an emotional one, one we still bear as a society as we acknowledge these traces of the past that remain with us today.

25

Riders of the Storms

Never before had I seen a flood tide actually flood. The water flowed up and over the paved road in front of us, a steady sheet not high enough to impede our vehicle but enough to make us stop, get out, and stare. The road itself was simple, straight, and narrow; it was one I had traveled many times to and from the University of Georgia Marine Institute and Nannygoat Beach on the south end of Sapelo Island. But seeing it immersed was a novel experience, especially knowing that the source of the water was from an ocean we could not yet see.

It was September 2015, and my occasion for paying yet another visit to Sapelo was not with a class, nor for research, but for the start of the *Georgia Coast Atlas*. As mentioned earlier, the atlas was (and still is) a multimedia project with colleagues from Emory University, in which we develop digital content—text, photos, videos, 360-degree panoramas, and more—and put these onto an Emory-hosted website so the general public and scholars alike can benefit from its educational value.[1] My friends who cowitnessed the drowning road were geographer and drone pilot Michael Page, his able field assistant (and wife) Rebecca Page, and videographers Steve Bransford and Anandi Salinas Knuppel. Our main goal during the three-day visit was to film short video clips with me explaining various aspects of the natural and human history of Sapelo, which later would be edited and added to our website. Additional goals included gathering aerial-drone footage of the various ecosystems on Sapelo, as well as time-lapse photography and panoramas. We fortuitously scheduled our three days of field time during a full moon, which ensured spring tides and the more spectacular visual effects they offered, especially for the time-lapse segments. For instance, imagine watching four hours of a rising tide on a Georgia coast beach condensed to less than a minute, its brevity conveying the sheer energy and volume of this phenomenon. In capturing this and other footage, we were successful. But we did not anticipate how the tides would exceed our expectations, as well as shorelines, tidal creeks, causeways, and more.

The drone footage Michael recorded during high tide on one of our days there was striking, eliciting wonder from us because of the pervasive effects of the tide. Even the "high marsh" areas of the salt marshes—including barren saltpans—were under water. All tidal creeks had overflowed and blurred the normally sharp boundaries of channels, banks, low marshes, high marshes, and so on. Instead, these distinct zones had merged into an ephemeral shallow-marine environment, their plants and animals buried under a vast layer of mocha-colored liquid. While viewing this video footage, I imagined millions of fiddler crabs in these ecosystems hunkering down and waiting in their burrows for the floodwaters to recede. I also thought of wading birds, raccoons, and other animals that typically fed on these crabs having to involuntarily eschew these crustaceans, rather than simply chew. Alligators, of course, would have been fine regardless, although any hanging out in dens near marshes surely had to move, as all usable air was displaced. Still, I also considered how the tracks and trails of alligators and other animals that used the marshes as walking routes were diminished by the inundation, their trace-making activities temporarily ceased.

Curious, we later read online news articles to see how human communities were affected by several days of flood tides.[2] While scanning these articles, we learned of a term that was mystifyingly new to me: "king tides." These tides have nothing to do with patrilineal royalty but instead fulfill their nicknames through absolute rule of coastal environments, including closing roads and isolating barrier-island communities. For instance, the causeway leading to and from Tybee Island was submerged during the high tides, temporarily stranding its residents. Once I read more research into the scientific explanation behind king tides, I found they were credited to a combination of spring tides, seasons, or nearby storms.[3] Yet we were only on the cusp of fall, and no storms were offshore during this time, meaning only a spring tide was involved.

Why were they so high, then? The answer came with the next spring tide in October 2015, as the same kind of flooding happened again. This time the people of Tybee Island were more prepared: as their paved link to the mainland went underwater again, they had planned accordingly to either leave or be stuck. Yet to have such an unusual event happen two consecutive months prompted a new conversation among people there and elsewhere on the Georgia coast. Public discussion of these king tides from the year before (2014) also included the words "sea-level rise" and "climate change," reflecting an increased awareness among residents on the coast of Georgia and other coasts throughout the world.[4]

As anyone with a modicum of evidence-based reasoning accepts, humans have changed the climate during the past few hundred years by doing the gaseous equivalent of placing a thick wool blanket over the earth's surface. Increased human production of carbon dioxide, methane, and other greenhouse gases, caused by our excessive burning of fossil fuels, has raised their amounts to their highest levels since the Pliocene epoch, which was about three million years ago.[5] The overall climate was accordingly warmer then, albeit not as warm as the last great "greenhouse Earth" of the Paleocene and Eocene epochs (about fifty-five to fifty million years ago), which enabled crocodiles and palm trees to live in Wyoming.[6] Following the warmth of the Pliocene epoch—which resulted in a barrier-island chain far up the present-day Georgia coastal plain—the Pleistocene epoch witnessed glacial and interglacial times. These cycles were driven mostly by astronomical causes, such as the tilt of the earth's axis and eccentricity of the earth's orbit, which in turn caused more or less solar heating of the earth's surface.[7] More warmth meant more melting of glacial ice, and, because water runs downhill in this reality, the oceans received a greater amount of water, which made the seas go up. Warmer oceans are also more expansive oceans, as greater temperatures contribute to greater volume.[8]

These Pleistocene fluctuations led to seesaw-like changes in the shoreline along the coastal plain, making and abandoning more barrier islands, their trace-making flora and fauna moving with each lateral shift. This is how fossil ghost-shrimp burrows ended up in sand ridges on the Lower Coastal Plain, and how the southernmost Georgia coast islands—Cumberland, Jekyll, St. Simons, Sapelo, St. Catherines, and Ossabaw—developed Pleistocene foundations in their western parts, marking the most recent sea-level high (and warmer time) from about forty thousand to fifty thousand years ago.[9] Fossil ghost-shrimp burrows in Pleistocene bedrock on the western edge of Sapelo likewise tell us the shoreline was there and that the soils underlying the Hog Hammock community to the east, which were later used to grow Sapelo sugarcane and red peas, were formed in part by waves and tides of shallow oceanic environments not so long ago.[10]

FIGURE 45. King tides as a harbinger of sea levels of the future. *Left*, flooded road on Sapelo Island, as the high tide passes over it from one marsh to another; photo taken on September 27, 2015. *Right*, aerial view of marshes on both sides of same road, with tidal creeks and their banks completely submerged; photo taken from still of drone video, shot by Michael Page, also on September 27, 2015.

Other incursions of the sea on inland environments of the Georgia barrier islands have become both more frequent and violent. A little more than
a year later, in December 2016, our atlas crew visited Ossabaw Island, where
we documented the aftermath of Hurricane Matthew there. Matthew was
a category 2 by the time it reached the Georgia coast on October 8 and was
the first hurricane to strike the coast in more than twenty years.[11] Because
Ossabaw is wide and mostly undeveloped, we saw what damage a hurricane
can inflict on such islands. As we moved along the sandy roads there, we
saw hundreds of large, old live oaks downed, oaks that had witnessed generations of artists, writers, and scientists coming to and leaving Ossabaw.
Their root systems were completely exposed to the air for the first and last
time; two months after faltering, their once vibrant evergreen leaves had
faded to dull browns. Also, our trips around the island were lengthened by a
breached east-west road, its modest causeway having been punched through
by floodwaters in a salt marsh and rendered impassable. We drove up to
the breach and, like on Sapelo with its king tide, got out of the vehicle and
stared in wonder at how change was happening, and we thought about how
we would deal with it.

Another hurricane-related experience the next year—in November
2017—made a third and final point for me that the Georgia coast had
reached a tipping point and was not turning back any time soon. I was on
the coast again, although not for *Georgia Coast Atlas* fieldwork. Instead, it
was my semiregular Thanksgiving break, in which my wife, Ruth, and I ride
bicycles on Jekyll Island and which, as you may recall, inevitably requires
stopping occasionally to admire the traces of rapacious gastropods and
shorebirds. Such frivolity, however, was interrupted on our first day there
when I discovered why my beach pedaling had become so laborious: I had a
flat tire. With no means on the island for repairing it, I was forced to rent a
bike the next day. Unfortunately, part of the rental agreement said it could
not be ridden on a beach unless I felt like paying an extra fifty dollars to
clean the bike. Given my thrifty ways, Ruth and I stayed on paved paths and
headed toward one of our favorite places on the island. It was a salt marsh in
the northeast side of the island, just west of an area nicknamed "Driftwood
Beach." This northern stretch of beach is so called because of its expanse of
dead, leafless trees in the surf there, marking the seaward edge of a maritime forest overwhelmed by the saltwater intrusion of a shifting shoreline.

It was while crossing the marsh on the elevated path and within sight of
Driftwood Beach that Hurricane Irma paused us: not the hurricane itself,

but its aftereffects. Part of the path was gone, an absence announced by a warning sign but clearly indicated by a 3-meter (10-foot) wide gap that, had we not minded it, surely would have resulted in more than fifty dollars of damages to the bicycle and me. A tidal creek running under the path had reclaimed this place, and a storm-washover fan of sand covered the marsh on the east side of the path that faced the open ocean. Tops of *Spartina* from the underlying marsh hinted of the environment there only three months before. Some of this *Spartina* was exposed by the tidal creek, which cut through the washover fan until dwindling just short of the beach, its former connection to the ocean severed. For the first time ever—on the west side of the path and well into the marsh—I saw sand ripples on the banks of normally mud-filled tidal creeks. Sand fiddler–crab burrows had perforated these ripples and the washover fan, replacing the burrows of mud fiddler crabs and other mud-loving tracemakers in this new substrate-controlled ecosystem. This once-familiar place had changed and radically so.

Hurricane Irma made landfall on the Georgia coast on September 11, 2017. The next day Ruth and I and about five million other people experienced Irma's ferocity far from the coast in the metropolitan Atlanta area as it flooded streets, ripped up old and cherished trees, knocked out power for hundreds of thousands, and killed.[12] It was a rare instance of a tropical storm impacting a major U.S. city that was more than 400 kilometers (250 miles) inland and at an elevation of about 300 meters (1,000 feet) above sea level.

On Jekyll Irma's powerful waves were even more dramatic, altering the north shore of the island and eroding what little land stood between the ocean and a new condominium development. The beach there was already diminished by longshore drift, which conspired with exotic boulders of gneiss (riprap) to wash away underlying sand. The rocky barrier, backed in places by a cement seawall, was placed there in a misguided attempt to stop beach erosion after Hurricane Dora hit Jekyll in 1964.[13] Nonetheless, the rocks starved the beach of sand, sand that really needed to accumulate around *Spartina* rack, which would have provided a foundation for sea oats, which in turn would have secured the sand with its roots and eventually made enough habitat for burrowing ghost crabs and nesting mother sea turtles. This rocky "fix" set up the shoreline for rapid undermining, as flooding in all coastal areas on Jekyll was worse than that caused by Hurricane Matthew the previous year. Irma also coincided with a high tide, which in turn led to greater volumes of water in this storm surge.[14] The surge left a

FIGURE 46. Storm damage on Ossabaw Island from Hurricane Matthew, which hit the Georgia coast on October 8, 2016. *Left*, toppled live oak, its roots exposed to light for the first time in more than a hundred years. *Right*, breached causeway, eroded by storm waves in the surrounding salt marsh. Photos taken December 12 and 13, 2016, respectively.

FIGURE 47. A storm-washover fan on the north end of Jekyll Island, composed of sand eroded from the beach and transported past what's left of a maritime forest, then dumped on top of what used to be a salt marsh. This meant that sand fiddler crabs began burrowing in a place previously occupied by mud fiddler crabs. Fan was caused by Hurricane Irma, which struck the southeastern United States on September 11, 2017; photo was taken on November 27, 2017.

narrower beach, a stark and isolated boulder field, a bluff, and a thin strip of land between these and the new condos. During a follow-up visit to Jekyll in January 2018, I saw that vanity had outweighed sanity, as giant white-canvas, cubic sandbags occupied the space between the condos and boulders. I later learned these were part of an intensive (and expensive) "rebuilding" effort, which will surely fail with the next few tropical storms.

The effects of Hurricane Irma were startling on other islands along the Georgia coast, again rearranging, literally overnight, the face of a place I had always taken for granted. On my next trip to Sapelo Island with a group of students, colleagues, and Ruth in February 2018, we did our usual walk through the varied ecosystems of the nature trail on the south end of the island. This was when we saw the effects of pine-beetle infestation (mentioned before), as well as trees downed by the storm. Other damage was subtler, but by the time we emerged from the trail onto Nannygoat Beach, I was already shaken by the changes, my illusion of familiarity denied. We continued on a long beach walk south, past a pavilion at the end of the same paved road I had seen beneath a king tide about eighteen months before. The area somehow looked different, but I could not yet say how.

As we walked, we saw some common items: ghost-crab burrows, tracks, and resting traces; hermit-crab trackways great and small; and shorebird tracks showing walking, landing, and taking off. Yet we also noticed oddities, seeming portents of altered times. For example, four turkey vultures (*Cathartes aura*), clustered around a large, dark lump on the beach, drew us in to investigate further. It was a headless shark, its severed head nowhere in sight, its body pockmarked by beak strikes, and the sand around it trampled by the vultures. We also saw feral-cattle tracks on the beach, only the second instance in all of my years visiting Sapelo. I figured these tracks belonged to a wandering lone bull, but I also wondered what compelled him to come to the shore and how he got there. Tall, coastal dunes on Sapelo normally present a barrier, effectively deterring an animal that big. Unfortunately, with the descent of the sun, I did not have time to backtrack him and learn more about his day, including how he accessed the beach so easily. But I got an inkling soon enough.

When it was time to walk back to the University of Georgia Marine Institute, we regrouped at the pavilion by the shore, and I climbed the steps of the boardwalk beside it and glanced back at the beach. My students, friends, and wife all heard the shock in my voice as I realized the source of my previous unease. The dunes that normally stood in front of the pavilion between it and the beach were gone. Wiped out. Erased. Vanished. Kaput. They were . . . well, you get the point. The storm surge from Irma had flattened the shoreline, having moved dune sands farther toward the interior of the island. Buried bushes of former back-dune meadows, denoted by twigs poking out of the sand, had become part of a new, transformed beach. I looked below the boardwalk, where I was normally guaranteed a view of ghost-crab burrows and tracks. None were there. And why should they be? Their habitat was somewhere else now.

Barrier islands are called "barriers" for a reason. These islands protect mainland environments, serving as a first line of defense and dampening the energy of storms as they literally dampen.[15] But unlike human-made walls or other structures, barrier islands do not stay fixed. Instead, their sediments shift, normally landward, with storms. With these storm-dominated movements, some plants and animals die. But many organisms have adapted to changes both sudden and gradual, leaving enough descendants to recolonize new environments that overlie older ones. Succession happens, and it leaves traces of this success. Tracks, burrows, nests, and other imprints of life push aside or otherwise mold these environments into ones resembling those before them, but just a little farther removed from where waves once washed shores.

The lessons provided by Sapelo, Jekyll, Ossabaw, and other Georgia coast islands after Hurricanes Matthew and Irma demonstrate how ecological and geologic processes merge once marine environments encroach on and engulf those places we called land. Yet the ultimate human trace in this intermingling is climate change, which is already leaving its mark on the Georgia barrier islands. This is a trace that could overwhelm and outlast shell rings, buildings, roads, and myriad other human modifications of the islands that we foolishly regard as fixed and permanent.

Given the effects of climate change looming over the Georgia coast and nearly everywhere else, we know to expect more king tides, more storms, and fiercer storms.[16] Nonetheless, we do not yet have a clear view of how quickly these lateral shifts might happen. To predict this uncertain future, we must look at and study plant and animal traces left after tides, hurricanes, and high sea levels of a prehuman past, better uniting our worldviews of before and after.

26

Vestiges of Future Coasts

The sea was getting closer. We had to work fast. I stood on a thinning beach between increasingly energetic ocean waves and a yellowish sandy bluff, three meters (ten feet) high, as the tide rose behind me. The water had begun to flow around the large, prone trees, their branches and trunks potentially impeding my escape. Fortunately, though, I was not alone, nor was I in mortal peril. For one, my ichnological colleague and good friend Andy Rindsberg of the University of West Alabama was with me. Second, although the outcrop was friable, it was also easily scalable. Last but not least, we were comfortably doing what geologists often do: standing in front of an outcrop and politely asking it to divulge previously untold secrets.

It was May 2007. Andy and I had visited this same outcrop on St. Catherines Island only two months before, on the first of a two-day Geological Society of America field trip that started and ended in Savannah.[1] The field trip was unusual in its leadership, having been organized and led by a consortium of paleontologists, geologists, and archaeologists. What they all shared, though, was an interest in St. Catherines and its complicated natural and human histories. These histories ranged from its prehuman Pleistocene past, to more recent Native American occupancy, to Spanish missionaries, to British and American plantations, to its modern role as a place for scientific research. It was a wonderful trip for its favorable weather and variety of field-oriented topics, all peopled by a mix of enthusiastic participants and leaders, no less. Among these leaders were longtime sea-turtle conservationist and paleontologist Gale Bishop; archaeologist David Hurst Thomas; coastal geologist V.J. "Jim" Henry; island manager Royce Hayes; and first-rate naturalist Carol Ruckdeschel, whom Andy and I met for the first time there.

As often happens on field trips, our outdoor exchanges the first day were both spontaneous and spirited, then our mutually experienced topics were reviewed and debated more earnestly that night, with at least some of that discussion fueled by adult libations. This socialization encouraged more open dialogues and enthusiastic questioning the next day, while also helping us become aware of what others knew or, in some cases, did not know. For example, I did not bother querying the archaeologically inclined folks about invertebrate trace fossils, and they quickly learned not to ask me anything about humans.

Nonetheless, my joyful enthusiasm about invertebrate trace fossils is what temporarily derailed the field trip that first day and ultimately compelled Andy and me to return to St. Catherines and this specific outcrop later in the year. On that scientifically fateful afternoon, Gale Bishop stopped us at Yellow Banks Bluff, so dubbed because of its yellowish sandy banks composing a modest bluff. At this stop Gale told us what was then known about this sedimentary deposit. We could see for ourselves it was composed of mostly softly bound and fine-grained sand stained light brown and yellow by clays and iron hydroxide, respectively. But we could also discern thin coffee-brown layers forming more resistant ledges on its otherwise near-vertical slope, lending it a steplike profile. We were told these darker layers were interpreted as marine firmgrounds, signifying offshore places and times when the sea level paused between rising and receding, with sediments somewhere between subjective categories of soft (squishy) and hard (solid). Firmgrounds attract certain types of invertebrates adapted for drilling into and living in them.[2] Accordingly, these substrates often host diagnostic suites of traces, and Andy and I hoped to find such trace fossils in these strata.

Sand filled our shoes as we walked onto the piles of loose, eroded sand at the base of the outcrop, slowing our investigation as we and about thirty other people slid one step back with every two steps forward. Tree roots further complicated our paleontological perusal as these pointy plant parts—emanating from a maritime forest above the bluff—protruded from the bluff. Because the sediments were originally deposited toward the last of the Pleistocene epoch (about twenty thousand to twenty-five thousand years ago), these roots reflected a temporal mixing of modern and ancient that reminded me of the relict marsh on Sapelo. Given such an outcrop, rock hammers and other standard geologic tools were inappropriate for peeking below its surface. Fortunately, the field trip leaders brought a few putty knives, allowing us to scrape away the loose sand covering the real outcrop and exposing vertical surfaces for us to examine properly.

That was when Andy and I first discovered what else about this outcrop reminded us of the relict marsh on Sapelo. The dark-brown layers were blurred or cut across by a blend of fossil root traces, insect burrows, and sand fiddler–crab burrows. The root trace fossils were unmistakable, tapering downward and forking in their bottommost extents. These traces alone told us we were looking at not marine firmgrounds but the remains of environments above the tides. The insect trace fossils were pencil-thick burrows bearing a series of darker curved lines in their sand-filled interiors. These meniscae showed where their makers—probably beetle larvae or cicada nymphs—pushed sand behind them with their legs and heads as they burrowed. These burrows were important environmental clues, as nearly all insects avoid marine environments, which restricts them to landward places. The sand fiddler–crab burrows, which had circular cross-sections and J-shaped profiles, also pointed to animals that were neither in maritime forests, nor in the lower part of a beach, but somewhere between. In short, we were looking at deposits made on the land but not quite in the sea. This novel revelation made for an interesting and fun exchange with other field-trip participants and ultimately led to Gale inviting us back to describe and interpret the outcrop more properly, which we did.

So just what were the original Pleistocene environments represented by the dark layers? On Andy's and my return visit, and after several days of intensively studying the lowermost part of the outcrop, we concluded these layers were made by storms, and, more specifically, they were storm-washover fans.[3] Moreover, we could see the evidence of ecological succession preserved there, from everyday quiescence, to sudden violence, to new opportunities for communities of animals and plants that ordinarily would not have lived there.

What preceded the storm layers and their colonizers? The sandy sediments beneath the dark layers contained probable ghost-crab burrows, suggesting they were living in dunes near the Pleistocene ocean. We attributed the coffee color of the darker layers (more Americano than espresso) to organics mixed in with the sand. Such accumulations are common on the Georgia coast today, evident as "coffee-grounds" of *Spartina* and other plants, their tissues milled by waves and tides and often filling ripple troughs on low-tide beaches.[4] But storm waves can carry these grounds past beaches, over berms and dunes, and onto land behind the dunes. The storm layers—thinner than this book but thicker than a magazine—implied these were the more distant edges of washover fans, with thicker parts closer to the shore.

After storm waves dumped their loads and the weather went back to normal, the fans presented fresh, smooth sandy surfaces seasoned by salty water and covering former ghost-crab territories. To nearby sand fiddler crabs that waited out the storms in their burrows, these deposits were the equivalent of billboards reading, "Find Your New Home Here!" So these crabs moved into their new neighborhoods, commenced burrowing, and promptly resumed the three F's of fiddler-crab life: feeding, fighting, and mating. Soon afterward (perhaps within just a few months) salt-tolerant plants would have colonized these environments in their own ways, their root traces adding to and reaching below the burrows. Fungi living around plant roots altered the chemical makeup of surrounding sand, their symbiosis enhancing root outlines. When fiddler crabs abandoned their burrows, these holes were filled by white windblown sand, neatly providing stark visual contrasts for ichnologists more than twenty thousand years later.[5] More sand brought by more wind would have further buried these burrows while also delivering seeds of plants more accustomed to terrestrial (low-salt) conditions. The growth of these plants in turn would have invited insects to join them, including those that spent part of their life cycles underground, such as cicada nymphs that made their living by sucking sap from roots. The extensive and pervasive back-filled burrows of these insects blurred root traces while also cutting across fiddler-crab burrows of generations past. Life would have continued until another storm pressed the ecological restart button, causing the same sequence of events to happen again and again.

It was remarkable enough that the stories of these Pleistocene storm strata and their interactions with coastal lives were preserved in Yellow Banks Bluff, telling us what happened then and who was there. More noteworthy, though, was their relative position to when Andy and I documented them in 2007. As we wrote and sketched in our field notebooks, the storm layers were between waist and chest heights for us. This was well above the then-current high tide, the latter marked by the modern beach beneath our feet, implying that the sea level was notably higher at that time in the Pleistocene. This in turn spoke of a warmer atmosphere, expanded oceans, and more common and more ferocious storms.

Knowing that the ancestors of modern sand fiddler crabs, plants, and insects shifted from one place to another through time in the context of a changing world made for a more compelling and poignant story, testifying to their resilience. We were also struck by knowing how storms of a prehuman past, but not so long ago (geologically speaking), told of "wiping a slate clean" and ecological succession that happened without human influence or interference, long before John Milton told of an ejection of our species from paradise.[6]

Still, a facet of Andy's and my documentation that still fills me with melancholic regret is our knowing that what we studied is gone. As the modern transgression unfolds and the sea level continues rising along the Georgia coast, the outward face of Yellow Banks Bluff is erased daily, undercut by tides, waves, and storm surges. Coastal geologists who have studied the rate of erosion on the northeastern edge of St. Catherines estimate a loss of 2–3 meters (6.7–10 feet) per year since the mid-nineteenth century.[7] Do the arithmetic, and by 2018 (eleven years after our field study), erosion would have caused a minimum of 22 meters (72 feet) of permanently lost land. This means the spot where Andy and I stood is now well offshore and underwater. Moreover, the once-living tree roots in the outcrop from the overlying maritime forest, which threatened to poke our eyes in 2007, are now dead, their owners having fallen into the surf and adding their bodies to the "tree boneyard" along that shore.

Part of science is repeatability. In geology we follow that spirit by placing our faith in outcrops as solid, static places that can be revisited by future generations of geologists, who then check and recheck the works of those before them. But destruction of the geologic record itself serves as a two-fold insult, expunging evidence as originally described and snatching away the chance to study it again. Nowadays technological optimists would probably tell Andy and me that we could have done a mosaic of digital photos to preserve a virtual record of the entire outcrop, or laser-scanned it, or used some other *Star Trek*–like method to ensure that this former part of the St. Catherines landscape could be reconstituted, sand grain by sand grain. Sadly, such hopes are about as realistic as transporting and reassembling the atoms of people from one place to another and expecting them to be the same people. The fact is that we do not have enough time, people, and access to the right tools to win this race against the sea, and anything we record digitally will just be ghosts in, well, machines.

Hurricanes Matthew (2016) and Irma (2017) were additional wake-up calls for coastal preservation, showing how formerly rare catastrophes made annual can hasten unprecedented rates of change for people living with the Georgia coast. These events, when combined with monthly king tides and incremental heightening of the sea, have lent to a foreboding sense of uncertainty among residents and visitors alike, their short-term anxieties joining intergenerational concern.

Still, despite Yellow Banks Bluff involuntarily donating its Pleistocene deposits to the sediment budget of the modern transgression, its lessons will outlive us. The deposits Andy and I studied recorded how plants and animals of the past adapted to brief moments of terror imbued by storms, as well as the gradual and inexorable effects of climate change. Because the Georgia barrier islands are unique in the eastern United States for holding evidence of Pleistocene sea levels, this meeting of past storm deposits with current storms accompanied by a rising sea also lent to my thinking about traces not yet made on the Georgia coast.

Imagining the Georgia barrier islands as "canaries in the coal mine" for climate change is probably too simplistic and inappropriate. After all, talking about canary mortality is more appropriate where they are native, rather than placing them far below ground and surrounding them with the buried remains of Carboniferous swamps. Instead, we might better reach for "ghost shrimp in the beach" as a guiding analogy. For one, as uncertain as we are about the fate of our species in the next hundred or thousand years, we can be sure that ghost shrimp living in their deep underground bunkers will survive whatever happens above. After all, their ancestors were burrowing during the Mesozoic era and kept passing on genes after a massive space rock smashed into the earth about sixty-six million years ago, wiping out all dinosaurs not lucky enough to be birds.[8] We also know ghost shrimp will react to the landward migration of beaches by simply moving with them, settling on shallow oceanic sands and digging in. The same will happen with many other species, including ghost crabs, polychaete worms, fiddler crabs, whelks, moon snails, and the ultimate survivors, horseshoe crabs. Their descendants will simply move with the shore, wherever that might be.

But a sea that rises too quickly and in ways that plants and animals did not previously experience in their deep-time lineages may also result in tragic losses. For example, sea turtles outlasted the dinosaurs, but their far lower numbers now leave them vulnerable, especially on sandy shores where their nests can be washed away or drowned by king tides and storms. Shorebirds are similarly exposed, threatened by the shrinking of already-limited seaside nesting grounds. A companion to climate change is the spread of invasive species, such as egg-eating feral hogs that further accelerate prenatal and premature deaths of turtles and birds, or the beetles that degrade maritime forests before ocean waters reach their roots. All these examples point to a crucial factor shared by all imperiled species, which is denying their ability to create another generation. As the sea goes up, we may need to say good-bye to some long-extant species.

Connected to shoreward environments are back-dune meadows and maritime forests, places the sea once wetted until the shoreline shifted elsewhere. But with sea-level rise, a basic geologic principle comes into play that helps us visualize what will happen to these places. The principle, which I teach to my introductory geology students, is nicknamed "Walther's Law," named after German geologist Johannes Walther, who first described it.[9] Granted, scientists are not so invested in surety to call a guiding principle a "law," as if it must be obeyed. Yet Walther's Law works well as a means for retrodicting (interpreting what happened in the past) as well as predicting (foretelling the future). Summarized as "laterally adjacent environments succeed one another vertically," it states that a vertical sequence of sedimentary strata inform geologists which environments were next to one another before one shifted over the other.[10]

Hence I feel fairly confident in saying that as continental ice sheets continue to melt and oceans expand over the next few hundred years, the subsequent sea-level rise will produce a vertical sequence of sediments and traces on the Georgia barrier islands, from bottom to top, like so:

- sand with abundant plant roots and insect burrows, including ant, bee, and wasp nests, as well as those of beetles and cicadas
- root traces from sea oats and creeping vines, accompanied by ghost-crab burrows and sea-turtle nests (but fewer of the latter)
- shorebird tracks, burrows made by moon snails, whelks, dwarf surf clams, and other molluscans
- ghost-shrimp burrows, stingray feeding pits, sea-cucumber burrows, and other signs of fully immersed marine environments

Maritime forests, denuded of redbay, pines, and other important plants that contribute to the foundations of these ecosystems, also do not promise

FIGURE 48. Art as hypothesis: a vertical sequence of traces predicted for the one hundred miles of the Georgia coast as sea level rises over the next one hundred years, in which traces of land-dwelling organisms are replaced by those of marine environments. Artwork titled *Abstractions of a Rising Sea* (2011), by Ruth Schowalter and Anthony J. Martin; watercolor on paper, 66 x 101 cm (26 x 40 in.).

"life as usual," and the traces of their demise would also become buried under deposits of an encroaching sea. The same would also happen with the few remaining freshwater wetlands, as windblown sands from a once-distant shore fill crayfish burrows and alligator dens. What is adjacent now will succeed in the future, placing vestiges on top of what was there before.

If one were to adopt a triage mentality by prioritizing preservation of specific Georgia coast environments in the face of climate change and sea-level rise, I vote for salt marshes first and all others next. Unlike most sea grasses, smooth cordgrass cannot live if immersed too often. Similarly, the periwinkles that normally graze on cordgrass leaves and stems will suffer from increased predation if forced to spend more time underwater, as fish and swimming crabs pick them off. Patchier distributions of *Spartina* and their periwinkle inhabitants would translate to far less organic material raining down on the marsh surfaces and fueling nutrient cycles there. Less *Spartina* also means fewer root systems holding the mud together and scarcer places for ribbed mussels and oysters to settle and deposit mud. Mud fiddler crabs would also not burrow in such environments, as their feeding depends on emergent mudflats. Hence the ecological services these crabs provide through mixing marsh soils and boosting nutrient cycling would cease, altering marshes both on and below their surfaces. Male fiddler claw-waving rituals are far less likely to be performed underwater, thus also decreasing their comedic potential. All in all, the marshes would no longer be marshes but open-water and opaque bays belying the poetry of Sidney Lanier and other artistic inspirations drawn from their beauty.

With climate change affecting Georgia marshes and other coastal environments, we can take two complementary actions: mitigation and adaptation. Mitigation means to do whatever we can to decrease emissions of greenhouse gases like carbon dioxide and methane. Forsaking fossil fuels as the foundation of our global economy is the most obvious solution to this problem, although changing personal lifestyles through collective action, as well as educating future generations, is also helpful. But with the feral cat out of the bag and carbon-dioxide levels already exceeding those of the warmest times of the Pleistocene epoch, adaptation is inevitably part of our future too. Adaptation (or acclimation) means adjusting to more common king tides and tropical storms as norms and saying no to building homes that would invite horseshoe crabs to lay eggs in living rooms and ghost shrimp to burrow in basements.[11]

In this dual strategy of mitigation and adaptation, we will also need to ensure preservation of the habitats for all species we hold dear while maintaining a can-do attitude and never losing our sense of wonder for what we have now. To encourage such optimism, we must remind ourselves that the traces of past plants and animals of the Georgia coast inform us of how they survived, shifting their homes with whatever nature threw at them, again and again. Even in a world without humans, species on the Georgia barrier islands will move with change, seeking newly formed environments as the sea goes up and over the places of the near-future past. If we look and listen carefully, vestiges of the past and present lives of our coast and their evocative stories can both advise and comfort, even as we prepare for whatever happens next.

APPENDIX *Georgia Coast Tracemakers*

Tracemakers are divided into plants and fungi, invertebrates, and vertebrates and listed alphabetically by common name.

A. Plants and Fungi

Common Name	Species Name
Blue-stain fungi	*Grosmannia clavigera*
Laurel wilt	*Raffaelea lauricola*
Live oak	*Quercus virginiana*
Redbay	*Persea borbonia*
Saltmeadow cordgrass	*Spartina patens*
Saw palmetto	*Serenoa repens*
Sea oats	*Uniola paniculata*
Sea-oxeye daisy	*Borrichia frutescens*
Smooth cordgrass	*Spartina alterniflora*
Spanish moss	*Tillandsia usneoides*

B. Invertebrates

Common Name	Species Name
Atlantic horseshoe crab	*Limulus polyphemus*
Atlantic oyster drill	*Urosalpinx cinerea*
Blue crab	*Callinectes sapidus*
Coquina clam	*Donax variabilis*

Dwarf surf clam	*Mulinia lateralis*
Eastern oyster	*Crassostrea virginica*
European green crab	*Carcinus maenas**
Ghost crab	*Ocypode quadrata*
Ghost shrimp (Carolinian)	*Callichirus major*
Ghost shrimp (Georgia)	*Biffarius biformis*
Horse-guard wasp	*Stictia carolina*
Knobbed whelk	*Busycon carica*
Marsh periwinkle	*Littoraria irrorata*
Moon snail	*Neverita duplicata*
Mud fiddler crab	*Uca pugnax*
Redbay ambrosia beetle	*Xyleborus glabratus**
Ribbed mussel	*Geukensia demissa*
Sand fiddler crab	*Uca pugilator*
Southern pine beetle	*Dendroctonus frontalis*
Southern quahog	*Mercenaria mercenaria*
Stone crab	*Menippe mercenaria*

C. Vertebrates

Common Name	Species Name
American alligator	*Alligator mississippiensis*
Bald eagle	*Haliaeetus leucocephalus*
Black bear	*Ursus americanus†*
Bobcat	*Lynx rufus*
Cat	*Felis catus**
Cattle	*Bos taurus**
Common ground dove	*Columbina passerina*
Cougar	*Puma concolor†*
Diamondback terrapin	*Malaclemys terrapin*
Eastern mole	*Scalopus aquaticus*
Great egret	*Ardea alba*
Green sea turtle	*Chelonia mydas*
Herring gull	*Larus argentatus*
Hog	*Sus scrofa**
Horse	*Equus caballus**
Laughing gull	*Leucophaeus altricilla*

Leatherback sea turtle	*Dermochelys coriacea*
Lionfish	*Pterois volitans**
Loggerhead sea turtle	*Caretta caretta*
Nine-banded armadillo	*Dasypus novemcinctus**
Raccoon	*Procyon lotor*
Red knot	*Calidris canutus*
Red wolf	*Canis rufus†*
Ring-billed gull	*Larus delawarensis*
River otter	*Lutra canadensis*
Sanderling	*Calidris alba*
Southeastern beach mouse	*Peromyscus polionotus*
Southern leopard frog	*Rana sphenocephala*
Southern toad	*Anaxyrus terrestris*
Spadefoot toad	*Scaphiopus holbrookii*
Star-nosed mole	*Condylura cristata*
Tricolored heron	*Egretta tricolor*
Turkey vulture	*Cathartes aura*
White-tailed deer	*Odocoileus virginianus*
Wolf	*Canis lupus*
Yellow-bellied sapsucker	*Sphyrapicus varius*

* Nonnative tracemakers
† Extirpated tracemakers

NOTES

Chapter 1. Knobbed Whelks, Dwarf Clams, and Shorebirds

1. Ruppert and Fox, 1988.

2. The shell of this large predatory marine snail was officially adopted as the Georgia state shell on April 16, 1987. Georgia state seashell, *Georgia Info*, https://georgiainfo.galileo.usg.edu/topics/government/article/georgia-state-symbols/georgia-state-seashell-knobbed-whelk.

3. Carriker, 1951; Peterson, 1982; Ruppert and Fox, 1988.

4. Dietl, 2004.

5. Trueman and Brown, 1992; Bromley, 1996.

6. Ruppert and Fox, 1988; B. Witherington and D. Witherington, 2011.

7. Martin, 2013.

8. Ibid.

9. Wilson, 2011.

10. Elbroch and Marks, 2001; Martin, 2013.

11. Martin, 2013.

12. Elbroch and Marks, 2001; Martin, 2013.

13. Prezant et al., 2002; Grant, 1981.

14. Howard and Elders, 1970.

Chapter 2. The Lost Barrier Islands of Georgia

1. Alligators belong to an evolutionarily linked group (clade) called Alligatoroidea, which originated in the Late Cretaceous period (more than sixty-six million years ago). In contrast, canids, which belong to the clade Canidae, date back only to the Eocene epoch, from about forty million years ago. Spaniels were not specially bred until probably the fourteenth century (one is mentioned by Chaucer in *The Canterbury Tales*) in Europe, which I might add has no extant crocodilians. Brochu, 2003; X. Wang and Tedford, 2008; Lloyd (1924), 2013.

2. Nichols, 2012.

3. Odum, 1968; Craige, 2002.

4. Martin, 2013.

5. W.A. Pirkle and F.L. Pirkle, 2007.

6. Given that there are so many introductory geology textbooks, narrowing it down to one is difficult. So if you do not already have one but want to learn some basic geology, I recommend an open-source textbook, licensed under Creative Commons, that can be downloaded free, such as *Physical Geology*, by Steven Earle at https://opentextbc.ca/geology/.

7. Nichols, 2012.

8. Ibid.

9. Martin, 2013.

10. Nichols, 2012.

11. Hoyt et al., 1964; Weimer and Hoyt, 1964; Hoyt and Hails, 1967; Hoyt and Henry, 1967.

12. Hoyt and Hails, 1967; Henry et al., 1993.

13. Hoyt and Hails, 1967; Linsley et al., 2008.

14. Uniformitarianism was also known as the Doctrine of Uniformity, and the concept was probably best described early on by geologist Charles Lyell (1797–1875), author of the classic work *Principles of Geology* (1830–1833). However, William Whewell (1794–1866) actually coined the term while describing Lyell's concept. He is also credited with the term "scientist," among others; see Hessenbruch, 2000.

15. Glasspool and A.C Scott, 2010.

16. Nel et al., 2018; Okajima, 2008; Falcon-Lang, 2000.

17. Sakai, 2005.

18. Schweigert, 2011.

19. Carmona et al., 2004; Gandini et al., 2010.

20. Martin, 2013.

21. Bromley, 1996 (and references therein).

22. Weimer and Hoyt, 1964.

Chapter 3. Georgia Salt Marshes, the Places with the Traces

1. Miami University in Oxford, Ohio, was chartered in 1809 and had its first students in 1824, whereas Florida did not become a U.S. state until 1845; see History and traditions, *About Miami*, https://miamioh.edu/about-miami /history-traditions/index.html; Statehood, *Florida Department of State*, https:// dos.myflorida.com/florida-facts/florida-history/a-brief-history/statehood/.

2. J.M. Teal, 1958; J.M. Teal, 1962; M. Teal and J.M. Teal, 1964.

3. B. Witherington and D. Witherington, 2011.

4. Sherr, 2015.

5. Martin, 2013 (and pertinent references therein).

6. J.M. Teal, 1962; M. Teal and J.M. Teal, 1964; Odum, 1968.

7. Martin, 2013.

8. "The Marshes of Glynn," written in 1875 by revered Georgia poet Sidney Lanier (1842–1881), is public domain. So you can read it in its entirety online (mint julep optional): *Poems*, https://www.poets.org/poetsorg/poem/marshes-glynn.

9. Potter et al., 1980.

10. J.M. Smith and Frey, 1985.

11. Fierstien and Rollins, 1987; Silliman and Bertness, 2002.

12. Martin, 2013.

13. J.M. Teal, 1958; J.M. Teal, 1962; Katz, 1980; J.Q. Wang et al., 2010.

Chapter 4. Rooted in Time

1. Martin et al., 2015; Dattilo et al., 2014.

2. Yefremov, 1940; Behrensmeyer et al., 2000.

3. Weigelt, 1927; also see the translation: Schaefer, 1989.

4. Rogers et al., 2010.

5. Frey and Howard, 1969; Frey and Basan, 1981; Morris and Rollins, 1977.

6. Martin, 2013 (and pertinent references therein).

7. Potter et al., 1980.

8. Proffitt et al., 2003.

9. Haas, 1950; Walker, 1989.

10. Martin, 2013.

11. J.M. Smith and Frey, 1985.

12. Mikuláš, 2001; Montalvo, 2002.

13. Gilman (1999), 2007.

Chapter 5. Coquina Clams, Listening to and Riding the Waves

1. B. Witherington and D. Witherington, 2011.

2. Martin, 2013.

3. Trueman et al., 1966.

4. Bromley, 1996; Martin, 2013.

5. Bromley, 1996; Pervesler and Uchman, 2009.

6. Maples and West, 1989; Ekdale and Bromley, 2001.

7. At the time of my writing this, the U.S. Geological Survey had a nicely summarized description of the Anastasia Formation, including primary references. I say "had" just in case the currently reigning executive branch erased it

and other scientific information since my writing this; see https://mrdata.usgs
.gov/geology/state/sgmc-unit.php?unit=FLPSa%3B0.

8. *Castillo de San Marcos*, 1993; Castillo de San Marcos National Monument, U.S.
National Park Service, https://www.nps.gov/casa/index.htm; Coquina: The rock
that saved St. Augustine, Fort Matanzas National Monument, U.S. National Park
Service, https://www.nps.gov/foma/learn/historyculture/upload/Coquina.pdf.

9. Raab, 2007.

10. Ellers, 1995a; Ellers, 1995b.

11. H.J. Turner and Belding, 1957.

Chapter 6. Ghost Crabs and Their Ghostly Traces

1. World Register of Marine Species, 1795.

2. Hill and Hunter, 1973; Frey et al., 1984; Duncan, 1986.

3. Wolcott, 1978; Robertson and Pfeiffer, 1982.

4. Greenaway et al., 1984; Farrely and Greenaway, 1994.

5. Negreiros-Fransozo et al., 2002; Subramoniam, 2016.

6. Hill and Hunter, 1973; Duncan, 1986; Bromley, 1996.

7. Martin, 2013.

8. Wolcott, 1984.

9. Martin, 2006b.

10. Pasini et al., 2016.

Chapter 7. Ghost Shrimp Whisperer

1. I could continue being a scholar and provide a number of peer-reviewed
articles about mantis shrimp, but we're talking about mantis shrimp, which
deserve way better than dry academic descriptions. So to get a proper introduction
to these incredible (and punchy) animals, here is a news article: Debczak, 2016.

2. Curran and Martin, 2003.

3. Bromley, 1996; Martin, 2013 (and references in both sources).

4. Van der Wal et al., 2017.

5. Sakai, 2005.

6. Weimer and Hoyt, 1964.

7. Carmona et al., 2004.

8. G.A. Bishop and E.C. Bishop, 1992; G.A. Bishop and Brannen, 1993.

Chapter 8. Why Horseshoe Crabs Are So Much Cooler Than Mermaids

1. Yes, I will gladly name and shame. The cable channel was Animal Planet, which produced *Mermaids: The Body Found* and *Mermaids: The New Evidence* and broadcast these shows in 2012 and 2013, respectively. Rather than risk brain-cell rot initiated by "content" associated with these fictional productions, you should instead read the 2016 essay by marine biologist and science communicator Andrew David Thaler: "The Politics of Fake Documentaries."

2. Rudkin et al., 2008; Lamsdell, 2015.

3. Shuster et al., 2003.

4. Ibid.

5. Fredericks, 2012.

6. Cramer, 2015.

7. Martin and Rindsberg, 2007; Rindsberg and Martin, 2015; Martin, 2013.

Chapter 9. Moon Snails and Necklaces of Death

1. Ruppert and Fox, 1988; B. Witherington and D. Witherington, 2011.

2. This ratio of foot to shell width (minimum 2:1) is based on my personal observations of moon snails with their foots fully extended. But I would appreciate others testing it with actual data.

3. Martin, 2013.

4. Bromley, 1996.

5. Carriker, 1981.

6. Kitching and Pearson, 1981.

7. Martin, 2013.

8. Carriker, 1981.

9. Ibid.; Kabat, 1990.

10. Visaggi et al., 2013.

11. An overview of brachiopods, including a mention of their importance in the fossil record and an extensive bibliography, is at Brachiopoda lamp-shells, Animal Diversity Web, https://animaldiversity.org/accounts/Brachiopoda/.

12. Harding et al., 2007.

13. J.M. Arnold and K.O. Arnold, 1969; Bromley, 1993.

14. Kelley and Hansen, 2001.

15. Vermeij, 1982.

16. Zipser and Vermeij, 1978; Turra et al., 2005.

17. Martin, 2013.

18. Walker, 1989; Walker, 1992.

19. Walker, 1989; Martin, 2013.

20. B. Witherington and D. Witherington, 2011.

21. A very good description of how moon-snail sand collars (egg cases) are formed, accompanied by clear illustrations and photos, is at the website *Moon Snail*, Reproduction Research Study 1. Tom Carefoot wrote it, and he's a retired University of British Columbia marine biologist; see http://www.asnailsodyssey .com/LEARNABOUT/MOLLUSCA/moonRepr.php.

22. Casey et al., 2016. (The authors point out that adult moon snails also eat a variety of foodstuffs.)

Chapter 10. Rising Seas and Étoufées

1. Longshaw and Stebbing, 2016.

2. Martin, 2013 (and references therein); Martin, 2017.

3. Skelton, 2010.

4. Kawai et al., 2015.

5. Breinholt et al., 2009.

6. Martin et al., 2008.

7. Ibid.

8. Veevers, 2006.

9. Martin, 2017.

10. Kawai et al., 2015.

11. McClain and Romaire, 2004.

12. An exception to this ecological "rule" of salinity intolerance in crayfish is the red swamp crayfish (*Procambarus clarkii*), which apparently can colonize brackish-water environments; see Bissattini et al., 2015.

13. Hobbs, 1981.

14. D. Smith et al., 2015.

15. Georgia Sea Turtle Center, Jekyll Island Authority, https://gstc.jekyllisland .com/.

Chapter 11. Burrowing Wasps and Baby Dinosaurs

1. Martin, 2013.

2. Berendt, 1994.

3. Landry et al., 2003.

4. Hooton et al., 2014; Gormally and Donovan, 2010.

5. The reason why I use quotation marks (or "scare quotes") around "renourishment" is because the sudden addition of massive volumes of sand to a beach system often causes many of the animals living in and above the sand to die, which is not so nourishing for them. Hence when viewed from an ecological

perspective, this term is oxymoronic. At the time of my writing this, the most recent beach renourishment project at Tybee Island was April 2018; see U.S. Army Corps of Engineers, 2018.

6. Evans and O'Neill, 2009.

7. Ibid.; Martin, 2013.

8. Evans and O'Neill, 2009.

9. Godfray, 1994.

10. Quicke, 2014.

11. Elbroch and Marks, 2001.

12. Because so many people have published peer-reviewed research articles about the geology and paleontology of the Two Medicine Formation, I can't narrow it down to just a few, although a few are cited later. But anyone wanting an introduction to it and its place in paleontological history could start with Horner and Gorman, 1988.

13. Horner and Makela, 1979; Horner, 1982; Horner, 1984.

14. Horner and Gorman, 1988.

15. Varricchio et al., 1999.

16. Horner and Gorman, 1988.

17. Martin and Varricchio, 2011.

18. Martin, 2013.

Chapter 12. Erasing the Tracks of a Monster

1. If you would rather read about the movie *Pacific Rim* (2013) than watch it, here's a well-done review by Matt Zoller Seitz at Roger Ebert's movie-review website, published online July 12, 2013, https://www.rogerebert.com/reviews /pacific-rim-2013.

2. O.H. Pilkey et al., 2017.

3. Halfpenny and Bruchac, 2002; Farlow et al., 2017.

4. Martin et al., 2015; Martin, 2017.

5. Martin, 2017.

6. Rosenblatt and Heithaus, 2015.

7. Martin, 2013.

8. Although dinosaurs remain the marquee stars of the Mesozoic era, crocodile-like animals were there with them, starting in the Triassic period at about 230 million years ago. For an introduction to the wild world of Triassic landscapes, I recommend starting with Sues and N. Fraser, 2010.

9. Martin et al., 2015; J. Reolid and M. Reolid, 2017.

Chapter 13. Traces of Toad Toiletry

1. For reasons far too complicated to explain here, Gale Bishop took down the site in September 2018, so it is no longer available online. He did, however, archive it so that posts can be retrieved from him on request, and I kept downloaded copies of his photos.
2. Halfpenny and Bruchac, 2002.
3. Elbroch, 2003.
4. Martin, 2013.
5. Ekdale and Bromley, 2001.
6. Rindsberg and Martin, 2003.

Chapter 14. Why Do Birds' Tracks Suddenly Appear?

1. Veevers, 2006.
2. Elbroch and Marks, 2001.
3. Lockley et al., 2015.
4. Pickerell, 2014.
5. Müller, 1956; Reineck, 1981.
6. Martin, 2013.
7. Ibid.
8. Since I first wrote about bird landings in 2013, the number of slow-motion videos of this behavior have happily increased and are easily found using keyword searches. Still, my favorite is this one of a sparrow both landing and taking off: *Sparrow Landing in UltraSlo Motion*, YouTube, March 20, 2019, https://www.youtube .com/watch?v=FoCMiRNksQo.
9. Martin, 2013.
10. Ibid.
11. Gladwell, 2005. For a more nuanced and in-depth analysis of the so-called thin-slicing (intuitive) thinking described by Gladwell, read Isenman, 2013.
12. Rich and Vickers-Rich, 2000.
13. Martin et al., 2012.
14. Martin et al., 2014.
15. Ibid.

Chapter 15. Traces of the Red Queen

1. Alexander and Henry, 2007.
2. Sobkowiak et al., 1989; Gerrard and Bortolotti, 2014.
3. Dial et al., 2006; Heers and Dial, 2012.

4. Van Valen, 1973; Van Valen, 1974.

5. Carroll, 1871.

Chapter 16. Marine Moles and Mistaken Science

1. Bradley, 2008.

2. Martin, 2013.

3. "Occam's razor" is a basic scientific principle in which one assumes the simplest explanation to a problem is probably the right one. This assumption, of course, also may be wrong. Yet it at least gives scientists a starting place when trying to disprove a series of hypotheses, and simpler hypotheses are often easier to test. It is named after thirteenth- to fourteenth-century English Franciscan friar and theological philosopher, William of Ockham. Interestingly (and perhaps not coincidentally), he is always depicted as clean shaven.

4. Frey and Pemberton, 1986.

5. Ruckdeschel, 2017.

6. Petersen and Yates, 1980; Gould et al., 1993.

7. Stomach contents for eastern moles have been studied, though, but not in a coastal setting; see Hartman, 1995.

8. If you are intrigued by schlocky 1950s science-fiction movies bad enough for lampooning in *Mystery Science Theater 3000*, then you're in for a treat with *The Mole People* (1956), which was directed by Virgil W. Vogel and stars no one you know. Even better, you can watch it for free, as the entire movie is archived online in the *Internet Archive*, https://archive.org/details/TheMolePeople.

9. Petersen and Yates, 1980.

10. Gingras et al., 2002; Martin, 2006a.

Chapter 17. Tracking That Is Otterly Delightful

1. Halfpenny and Bruchac, 2002; Elbroch, 2003.

Chapter 18. Alien Invaders of the Georgia Coast

1. Mann, 2011.

2. Anthony, 2017.

3. At the time of my writing this book, Australia had no definitive evidence of dingoes living there until about 3,500 years ago; see B. Smith and Savolainen, 2015.

4. The most well known of indigenous dingo legends was of a massive dingo named Gaiya who was controlled by Eelgin the Grasshopper Woman and used to

kill people, but later two brothers convert him into two "good" dingoes. The story was written and illustrated in a children's book, *The Giant Devil Dingo* (1973), by Dick Roughsey (Angus and Robertson Children's Books, Sydney, Australia: 34 p.). Weston Woods Studio produced a video with a reading of the story combined with the book's images in 1980. Also be sure to look for tracks and tracking as part of the story. *Internet Archive*, https://archive.org/details/thegiantdevildingo.

5. Letnic, 2012.

6. Hoffecker, 2017.

7. Anthony, 2017.

8. The University of Georgia Marine Extension has an informative website about aquatic invasive species in Georgia, appropriately titled "Aquatic invasive species," with links to fact sheets: https://gacoast.uga.edu/outreach/programs/aquatic -invasive-species/. I also recommend looking at "Species of Concern," Georgia Invasive Species Task Force site, hosted by the Center for Invasive Species and Ecosystem Health at the University of Georgia, https://www.gainvasives.org/species/.

9. Humprey, 1974; Sherr, 2015.

Chapter 19. The Wild Cattle of Sapelo

1. The largest mammals that could have been on any of the Georgia barrier islands during the Pleistocene include mastodons and mammoths (*Mammut* and *Mammuthus*, respectively), giant ground sloths (*Eremotherium*, *Paramylodon*), and bison (*Bison*). But we know surprisingly little about Pleistocene megafauna living on the present-day Georgia barrier islands, as few people have studied vertebrate fossils from islands with Pleistocene bedrock. This paucity is either a function of rare fossils, not enough people studying them, lack of funding, or a combination of these factors. Two now-dated peer-reviewed reports on Pleistocene vertebrates from coastal Georgia and Sapelo Island (respectively) are Hulbert and Pratt, 1998; and Laerm et al., 1999.

2. *Georgia Coast Atlas*, Emory University, www.georgiacoastatlas.org.

3. The video might be a little difficult to find on the *Georgia Coast Atlas* site, but it is one of several tagged on the digital map of Sapelo at its south end, near the intersection of Horse Pasture Road and Greenhouse Way. Look for one with me on one knee in front of old feral-cattle scat: *Sapelo Island: Tracking Animals*, Emory Center for Digital Scholarship, http://georgiacoastatlas.org/sapelo-island.html.

4. Thompson and Turck, 2010.

5. I could not find any references to feral cattle on other barrier islands in the United States, which makes Sapelo unique in this respect. But Cumberland Island had some feral cattle starting in the first half of the twentieth century and until 1987, which Carol Ruckdeschel discusses in her book (2017).

6. C. Bailey and Bledsoe, 2000; Cooper, 2017. If you are interested in learning more about R.J. Reynolds Jr. and his tobacco-fueled life, there's also this book: Schnakenberg, 2010.

7. Stuart, 2017.

8. Ajmone-Marsan, 2010; McTavish et al., 2013.

9. Rokosz, 1995; Van Vuure, 2005. Julius Caesar witnessed aurochs during his campaign in Gaul, and, judging from the following passage, he was quite impressed by them: "There is a third kind, consisting of those animals which are called *uri* [aurochs]. These are a little below the elephant in size, and of the appearance, color, and shape of a bull. Their strength and speed are extraordinary; they spare neither man nor wild beast which they have espied." He wrote this in *Commentarii de Bello Gallico*, or *De Bello Gallico* (The Gallic War), which is available online through Project Gutenberg: http://www.gutenberg.org/ebooks/10657, bk. 6, p. 28.

10. I couldn't find the location of Harder's original painting, nor whether it might be viewed publicly, but an online image was on Wikipedia Commons: *A Long-Horned European Wild Ox Attacked by Wolves*, https://upload.wikimedia.org /wikipedia/commons/2/28/Long_horned_european_wild_ox.jpg.

11. If you need to know more about this bovine love story (actually, a tale of wooing and rejection), all told through their tracks, I describe it in detail in my book *Life Traces of the Georgia Coast* (2013) on page 432.

12. Martin, 2013.

13. I couldn't find primary historical references about the feeding of smooth cordgrass to cattle, but in my search I found that herds of wild cattle lived on coastal Georgia during the eighteenth century and ate marsh vegetation there; see Stewart, 2002. Also, Dr. Evelyn Sherr, a longtime resident of Sapelo and marine biologist at the University of Georgia Marine Institute, told me that New Englanders used the closely related salt-meadow cordgrass (*Spartina patens*) as food for cattle. This plant is also common along the edges of Georgia salt marshes and in beach-dune meadows.

14. Hulbert and Pratt, 1998; Laerm et al., 1998.

15. Lott, 2002; Cunfer and Waiser, 2016.

16. Van Vuure, 2005; Van Vuure, 2014.

Chapter 20. Your Cumberland Island Pony, Neither Friend nor Magic

1. *Foundation Document: Cumberland Island National Seashore*, February 2014. National Park Service, U.S. Department of the Interior, Washington, D.C.: 63 p., https://www.nps.gov/cuis/learn/management/upload/CUIS_FD_FINAL.pdf.

2. Ruckdeschel, 2017.

3. Yes, Spanish colonizers brought horses to North America during the fifteenth to seventeenth centuries, but there is no evidence that the horses on Cumberland Island are directly descended from that stock. So please stop saying that.

4. Ruckdeschel, 2017; Goodloe et al., 2000; Dilsaver, 2004.

5. Hulbert and Pratt, 1998.

6. Barrón-Ortiz et al., 2017.

7. Hulbert and Pratt, 1998; Laerm et al., 1999.

8. Elbroch, 2003; Martin, 2013.

9. Ibid.

10. At the time of my writing this book in late 2018 to early 2019, I was directing an undergraduate research student at Emory University (Arbour Guthrie) to map these trails in the southeastern part of Cumberland Island using geographic information systems. I am happy to state she successfully mapped the trails and showed how from 1990 to 2018 they increased significantly in both number and length, which in turn altered ecosystems in that part of the island. An abstract summarizing her research is Guthrie et al., 2019. Even better, Arbour graduated from Emory with Highest Honors in May 2019.

11. Bellis and Keough, 1995.

12. Browse lines caused by overbrowsing of herbivores are subtle traces, but once recognized are easily recognized. For more information, read Elbroch, 2003; and Martin, 2013.

13. M.G. Turner, 1987; M.G. Turner, 1988; Ruckdeschel, 2017.

14. Luther and Benzing, 2009.

15. Cumberland Island does have native dung beetles, which never have to worry about going hungry or feeding their children; see Fincher and R.E. Woodruff, 1979.

16. Martin, 2013.

17. Buynevich et al., 2011.

18. Chin, 2007; Chin, 2012.

19. De Stoppelaire et al., 2004.

20. Barber and P.H. Pilkey, 2001.

21. Sabine et al., 2006.

Chapter 21. Going Hog Wild on the Georgia Coast

1. Alexander and Henry, 2007.

2. W.B. Johnson, 2002.

3. W.J. Fraser, 2006.

4. Clayton et al., 1992.

5. Wood and Roark, 1980; Mayor and Brisbin, 2008; Ballari and Barrios-García, 2014.

6. Ruckdeschel and Shoop, 2012; Engeman et al., 2016.

7. Ruckdeschel and Shoop, 2012.

8. Hackney et al., 2013; Buzuleciu et al., 2016.

9. Brennessel, 2006; Quammen, 1996. Brennessel, 2006, includes the original White House recipe for turtle soup as enjoyed by U.S. president Howard Taft. I urge anyone interested in trying this recipe to find a substitute for terrapin as a main ingredient. How about pork?

10. Hume, 2012; Rijsdijk et al., 2015; Fuller, 2002.

11. Nichols, 2012.

12. Mayor and Brisbin, 2008.

13. Giuffra et al., 2000.

14. G. Larson et al., 2008.

15. Collared peccaries are relatively small compared with most domestic hogs, weighing in at a maximum of about thirty kilograms (forty-five pounds). In contrast, one of the collared peccarry's Pleistocene ancestors, the long-nosed peccary, may have been at least twice as heavy. A good summary of collared peccaries, or javelina, with references to further reading, is at the *Animal Diversity Web* site: Museum of Zoology, University of Michigan, https://animaldiversity .org/accounts/Pecari_tajacu/. I also recommend reading this nicely written and researched blog post from July 12, 2014, by paleontologist Aaron Woodruff, about the Pleistocene and Holocene history of peccaries in the Americas: "America's original razorbacks." The post also includes his fine artwork depicting these extinct animals.

16. Mayer and Brisbin, 2008.

17. Burgos-Paz et al., 2013.

18. R.B. Taylor et al., 1998.

19. Held et al., 2005; Broom et al., 2009; Kouwenberg et al., 2009.

20. Elbroch, 2003.

21. Sharp and Angelini, 2016.

22. Pejchar and Mooney, 2009.

23. Barrios-Garcia and Ballari, 2012.

24. Baker et al., 2001.

25. Beeland, 2013.

Ch*apter 22.* Redbays and Ambrosia Beetles

1. Rabaglia et al., 2006.

2. Cameron et al., 2008; Zomlefer et al., 2008.

3. Spiegel and Leege, 2013.

4. Coder, 2012.

5. Vega and Hofstetter, 2014.

6. Brar et al., 2013.

7. Harrington et al., 2008.

8. Vega and Hofstetter, 2014.

9. Lieutier et al., 2007; Braiser, 2012.

10. Martin, 2013.

11. Vega and Hofstetter, 2014.

12. Tapanila and Roberts, 2012.

13. Chin, 2007.

14. Labandeira, 2013.

15. Legendre et al., 2015.

16. Peris et al., 2014; Rehan, 2012.

17. S.E. Smith and Read, 2008.

18. J.W. Taylor et al., 2004.

19. Harrington et al., 2011.

20. Coder, 2012.

21. Mayfield et al., 2008; Kendra et al., 2013.

22. Meeker et al., 1995.

23. Bentz and Jönsson, 2014.

Chapter 23. Shell Rings and Tabby Ruins

1. C. Bailey and Bledsoe, 2000; M.N. Johnson, 2009; Cooper, 2017.

2. L.H. Larson, 1991.

3. Manucy, 1952; Sickels-Taves and Sheehan, 1999. Also noteworthy is that Muslims who spoke Arabic were among enslaved Africans in the Americas, and one of the founding members of the African population on Sapelo was Bilali, who was an ancestor of Cornelia Bailey (C. Baily and Bledsoe, 2000). Hence "tabby" may have had a more direct origin from Arabic in coastal Georgia.

4. Sickels-Taves and Sheehan, 2002.

5. Thompson, 2007.

6. Marquardt, 2010.

7. Thompson et al., 2004.

8. Russo, 2006; Turck and Thompson, 2016.

9. Worth, 2007.

10. Milanich, 1996.

11. Russo, 2006; Turck and Thompson, 2016.

12. Sassaman, 1993.

13. Thompson, 2007.

14. M.E. Sanger, 2015.

15. Andrus and Thompson, 2012.

16. Beeland, 2013; Hornocker and Negri, 2009.

17. Ruckdeschel, 2017.

18. Davis, 2011.

19. C.K. Smith and McGrath, 2011; Vanderplank et al., 2014.

20. M.C. Sanger and Thomas, 2010; M.E. Sanger, 2015.

21. Sickels-Taves and Sheehan, 1999.

22. Admittedly, I couldn't find a specific reference documenting which shell middens were mined to make tabby structures at Fort Frederica and Cannon Point on St. Simons Island, let alone a written directive by General Oglethorpe definitively stating, "Hear ye, hear ye, let us dig yon Indian shell mounds" or similar sentiments. But locally derived concentrated shell deposits for such extensive building would have been most easily supplied by previous generations of Native Americans rather than by colonists harvesting millions of oyster shells from local marshes. Anyway, for more information about Fort Frederica and its place in Georgia-coast history, there is J.T. Scott, 1994.

Chapter 24. Ballast of the Past

1. Burstöm, 2018.

2. Stewart, 2002; Harris and Berry, 2014.

3. O. Pilkey and M.E. Fraser, 2003.

4. Hess et al., 2002.

5. Dahanayake et al., 1985.

6. Martin et al., 2015.

7. Martin, 2013.

8. The largest slave auction in the history of the United States was in Savannah on March 2–3, 1859 (only two years before the U.S. Civil War), in which plantation owner Pierce M. Butler sold 436 men, women, and children to the highest bidders. African Americans later referred to it as "The Weeping Time" because of the heavy rain that fell throughout the auction, symbolizing tears from above. In 2014 Kristopher Monroe wrote a synopsis of this event. For a more in-depth account, read A.C. Bailey, 2017.

9. Harris and Berry, 2014.

Chapter 25. Riders of the Storms

1. *Georgia Coast Atlas*, Emory University, www.georgiacoastatlas.org.

2. Galloway, 2015; Landers, 2015.

3. As of May 2019, the National Oceanographic and Atmospheric Administration still had a succinct explanation of king tides on its website; see What is a king tide?, National Ocean Service, https://oceanservice.noaa.gov/facts/kingtide.html.

4. Landers, 2014.

5. Brighan-Grette et al., 2013; Ogburn, 2013.

6. Paleoclimatologists refer to the most recent epically warm time in the earth's history as the Paleocene-Eocene thermal maximum, which was about fifty-five to fifty million years ago. A good scientific overview of the PETM and its applications to understanding modern climate change is McInerney and Wing, 2011. For a more general (but still informative) article, read Colin, 2015.

7. Astronomically driven cycles in climate, which depend on combinations of axial tilt, precession, and orbital eccentricity of the earth around the sun, are called Milankovitch cycles, named after their discoverer, Serbian science polymath Milutin Milanković (1879–1958). These regular cycles happened (and will continue to happen) as a result of the earth's position relative to the sun. If my words here fail to explain this adequately, the University of Wisconsin has more words and an interactive visualization here: The Vostok core and Milankovitch cycles climate applet, https://cimss.ssec.wisc.edu/wxfest/Milankovitch/earthorbit.html.

8. Warmer ocean water corresponds to less dense (expanded) ocean water, which, along with volume, increases relative sea level. The Intergovernmental Panel on Climate Change has a more detailed explanation of this phenomenon here: *Global Warming of 1.5 °C*, https://www.ipcc.ch/sr15/.

9. Linsley et al., 2008.

10. One example of the cultural revivification of Gullah-Geechee knowledge happening on Sapelo Island is the bringing back of heirloom breeds of sugarcane and red peas originally grown there; see Dixon, 2017; and Locker, 2017.

11. MacDonald, 2016.

12. Issa et al., 2018.

13. Clayton et al., 1992; Yang et al., 2012.

14. Bacopoulos, 2018.

15. O. Pilkey and M.E. Fraser, 2003.

16. Jisan et al., 2018.

Chapter 26. Vestiges of Future Coasts

1. G.A. Bishop et al., 2007.
2. Buatois and Mángano, 2011.
3. Martin and Rindsberg, 2011.
4. Martin, 2013.
5. Martin and Rindsberg, 2011.
6. *Paradise Lost*, an epic poem by John Milton (1609–1674) about the loss of innocence for humankind represented by the Genesis story, was first published in 1667 and remains one of the greatest works of English literature. The 1674 version is available through twenty-first technology here; see Poetry Foundation, https://www.poetryfoundation.org/poems/45718/paradise-lost-book-1-1674-version.
7. G.A. Bishop and Meyer, 2011.
8. Martin, 2017.
9. Seibold and Berger, 2017.
10. Nichols, 2012, 83.
11. Zannuttigh et al., 2014.

REFERENCES

Ajmone-Marsan, P., et al. 2010. On the origin of cattle: How aurochs became cattle and colonized the world. *Evolutionary Anthropology*, 19: 148–157.

Alexander, C., and Henry, V.J., Jr. 2007. Wassaw and Tybee Islands: Comparing undeveloped and developed barrier islands. In: Rich, F.J. (editor), *Guide to Fieldtrips, 56th Annual Meeting, Southeastern Section of the Geological Society of America*. Department of Geology and Geography Contribution Series, Georgia Southern University, 1: 187–198.

Andrus, C.F.T., and Thompson, V.D. 2012. Determining the habitats of mollusk collection at the Sapelo Island shell ring complex, Georgia, USA using oxygen isotope sclerochronology. *Journal of Archaeological Science*, 39: 215–228.

Anthony, L. 2017. *The Aliens among Us: How Invasive Species Are Transforming the Planet and Ourselves*. Yale University Press. New Haven, Conn.: 400 p.

Arnold, J.M., and Arnold, K.O. 1969. Some aspects of hole-boring predation by *Octopus vulgaris*. *American Zoologist*, 9: 991–996.

Bacopoulos, P. 2018. Extreme low and high waters due to a large and powerful tropical cyclone: Hurricane Irma (2017). *Natural Hazards*, 94: 1–30. https://doi.org/10.1007/s11069-018-3327-7.

Bailey, A.C. 2017. *The Weeping Time: Memory and the Largest Slave Auction in American History*. Cambridge University Press, New York: 206 p.

Bailey, C., and Bledsoe, C. 2000. *God, Dr. Buzzard, and the Bolito Man: A Saltwater Geechee Talks about Life*. Doubleday, New York: 334 p.

Baker, L.A., et al. 2001. Prey selection by reintroduced bobcats (*Lynx rufus*) on Cumberland Island, Georgia. *American Midland Naturalist*, 145: 80–93.

Ballari, S.A., and Barrios-García, M.N. 2014. A review of wild boar *Sus scrofa* diet and factors affecting food selection in native and introduced ranges. *Mammal Review*, 44: 124–134.

Barber, D.C., and Pilkey, P.H. 2001. Influence of grazing on barrier island vegetation and geomorphology, coastal North Carolina. *Geological Society of*

America Abstracts with Programs, 166. November 6. https://gsa.confex.com
/gsa/2001AM/finalprogram/abstract_28327.htm.

Barras, C. 2015. When global warming made our Earth super hot. BBC *Earth*,
September 14. http://www.bbc.com/earth/story/20150914-when-global
-warming-made-our-world-super-hot.

Barrios-Garcia, M.N., and Ballari, S.A. 2012. Impact of wild boar (*Sus scrofa*) in its
introduced and native range: A review. *Biological Invasions*, 14: 2283–2300.

Barrón-Ortiz, C.I., et al. 2017. Cheek tooth morphology and ancient mitochondrial
DNA of late Pleistocene horses from the western interior of North America:
Implications for the taxonomy of North American Late Pleistocene *Equus*. PLOS
ONE 12: e0183045. https://doi.org/10.1371/journal.pone.0183045.

Beeland, T.D. 2013. *The Secret World of Red Wolves: The Fight to Save North America's
Other Wolf.* University of North Carolina Press, Chapel Hill: 272 p.

Behrensmeyer, A.K., et al. 2000. Taphonomy and paleobiology. *Paleobiology*, 26:
103–147.

Bellis, V.J., and Keough, J.R. 1995. *Ecology of the Maritime Forests of the Southern
Atlantic Coast: A Community Profile.* Biological Report 30. National Biological
Service, U.S. Department of the Interior, Washington, D.C.: 95 p.

Bentz, B.J., and Jönsson, A.M. 2014. Modeling bark beetle responses to climate
change. In: Vega, F.E., and Hofstetter, R.W. (editors), *Bark Beetles: Biology and
Ecology of Native and Invasive Species.* Academic Press, Cambridge, Mass.: 533–553.

Berendt, J. 1994. *Midnight in the Garden of Good and Evil.* Random House, New York:
400 p.

Bishop, G.A., and Bishop, E.C. 1992. Distribution of ghost shrimp, North Beach, St.
Catherines Island. *American Museum Novitates*, 3042: 1–17.

Bishop, G.A., and Brannen, N.A. 1993. Ecology and paleoecology of Georgia ghost
shrimp. In: Farrell, K.M., et al. (editors), *Geomorphology and Facies Relationships
of Quaternary Barrier Island Complexes Near St. Mary's Georgia. Georgia Geological
Society Guidebook*, 13: 19–29.

Bishop, G.A., and Meyer, B.K. 2011. Sea turtle habitat deterioration on St.
Catherines Island: Defining the modern transgression. In: Bishop, G.A., et al.
(editors), *Geoarchaeology of St. Catherines Island, Georgia.* Anthropological Papers
of the American Museum of Natural History, no. 94: 271–295.

Bishop, G.A., et al. 2007. Transgressive barrier island features of St. Catherines
Island, Georgia. In: Rich, F.J. (editor), *Guide to Fieldtrips, 56th Annual Meeting,
Southeastern Section of the Geological Society of America.* Georgia Southern
University, Department of Geology and Geography Contribution Series, no. 1:
39–85.

Bissattini, A.M., et al. 2015. Tolerance of increasing water salinity in the red swamp
crayfish *Procambarus clarkii. Journal of Crustacean Biology*, 35: 682–685.

Bradley, T.J. 2008. Saline-water insects: Ecology, physiology and evolution. In: Lancaster, J., and Briers, R.A. (editors), *Aquatic Insects: Challenges to Populations*. Proceedings of the Royal Entomological Society's Twenty-Fourth Symposium, CAB International, Oxfordshire, U.K.: 20–35.

Braiser, C.M. 2012. Intercontinental spread and continuing evolution of the Dutch elm disease pathogens. In: Dunn, C.P. (editor), *The Elms: Breeding, Conservation, and Disease Management*. Springer, Berlin: 61–72.

Brar, G.S., et al. 2013. Life cycle, development, and culture of *Xyleborus glabratus* (Coleoptera: Curculionidae: Scolytinae). *Florida Entomologist*, 96: 1158–1167.

Breinholt, E., et al. 2009. The timing and diversification of the freshwater crayfish. In: Martin, J.W., et al. (editors), *Decapod Crustacean Phyogenetics*. CRC, Boca Raton, Fla.: 343–356.

Brennessel, B. 2006. *Diamonds in the Marsh: A Natural History of the Diamondback Terrapin*. University Press of New England, Lebanon, N.H.: 219 p.

Brighan-Grette, J., et al. 2013. Pliocene warmth, polar amplification, and stepped Pleistocene cooling recorded in NE Arctic Russia. *Science*, 340: 1421–1427.

Brochu, C.A. 2003. Phylogenetic approaches toward crocodylian history. *Annual Review of Earth and Planetary Sciences*, 31: 357–397.

Bromley, R.G. 1993. Predation habits of octopus past and present and a new ichnospecies, *Oichnus ovalis*. *Geological Society of Denmark Bulletin*, 40: 167–173.

Bromley, R.G. 1996. *Trace Fossils: Biology and Taphonomy*. Routledge, New York: 384 p.

Broom, D.M., et al. 2009. Pigs learn what a mirror image represents and use it to obtain information. *Animal Behavior*, 78: 1037–1041.

Buatois, L.A., and Mángano, M.G. 2011. *Ichnology: Organism-Substrate Interactions in Space and Time*. Cambridge University Press, New York: 371 p.

Burgos-Paz, W., et al. 2013. Porcine colonization of the Americas: A 60k SNP story. *Heredity*, 110: 321–330.

Burstöm, M. 2018. *Ballast: Laden with History*. Nordic Academic Press, Lund, Sweden: 120 p.

Buynevich, I.V., et al. 2011. Ungulate tracks in coastal sands: Recognition and sedimentological significance. *Journal of Coastal Research*, 64: 334–338.

Buzuleciu, S.A., et al. 2016. Scent of disinterred soil as an olfactory cue used by raccoons to locate nests of diamond-backed terrapins (*Malaclemys terrapin*). *Herpetological Conservation and Biology*, 11: 539–551.

Cameron, R.S., et al. 2008. *Distribution and Spread of Laurel Wilt Disease in Georgia: 2006–08 Survey and Field Observations*. Georgia Forestry Commission, Dry Branch: 29 p.

Carmona, N.B., et al. 2004. The trace fossil record of burrowing decapod crustaceans: Evaluating evolutionary radiations and behavioural convergence. *Fossils and Strata*, 51: 141–153.

Carriker, M.R. 1951. Observations on the penetration of tightly closing bivalves by *Busycon* and other predators. *Ecology*, 32: 73–83.

Carriker, M.R. 1981. Shell penetration and feeding by naticacean and muricacean predatory gastropods: A synthesis. *Malacologia*, 20: 403–422.

Carroll, L. 1865. *Alice's Adventures in Wonderland*. London, Macmillan.

Carroll, L. (or Dodgson, C.L.). 1871. *Through the Looking-Glass, and What Alice Found There*. Altemus, Philadelphia.

Casey, M.M., et al. 2016. You are what you eat: Stable isotopic evidence indicates that the naticid gastropod *Neverita duplicata* is an omnivore. *Frontiers of Ecology and Evolution*, 4: 125. https://doi.org/10.3389/fevo.2016.00125.

Castillo de San Marcos: A Guide to the Castillo de San Marcos National Monument, Florida. 1993. U.S. National Park Service, Government Printing Office, Washington, D.C.: 63 p.

Chin, K. 2007. The paleobiological implications of herbivorous dinosaur coprolites from the Upper Cretaceous Two Medicine Formation of Montana: Why eat wood? *Palaios*, 22: 554–566.

Chin, K. 2012. What did dinosaurs eat: Coprolites and other direct evidence of dinosaur diets. In: Brett-Surman, M.K., et al. (editors), *The Complete Dinosaur* (2nd Edition). Indiana University Press, Bloomington: 589–601.

Clayton, T.D., et al. 1992. *Living with the Georgia Shore*. Duke University Press, Durham: 188 p.

Coder, K.D. 2012. Redbay (*Persea borbonia*): drifting toward oblivion. Warnell School of Forestry and Natural Resources Native Tree Series. WSFNR12–9. Athens: 17 p.

Cooper, M.L. 2017. *Making Gullah: A History of Sapelo Islanders, Race, and the American Imagination*. University of North Carolina Press, Chapel Hill: 304 p.

Craige, B.J. 2002. *Eugene Odum: Ecosystem Ecologist and Environmentalist*. University of Georgia Press, Athens: 226 p.

Cramer, D. 2015. *The Narrow Edge: A Tiny Bird, an Ancient Crab, and an Epic Journey*. Yale University Press, New Haven: 293 p.

Cunfer, G., and Waiser, B. 2016. *Bison and People on the North American Great Plains: A Deep Environmental History*. Texas A&M Press, Lubbock: 344 p.

Curran, H.A., and Martin, A.J. 2003. Intertidal mounds of tropical callianassids provide substrates for complex upogebiid shrimp burrows: Modern and Pleistocene examples from the Bahamas. *Palaeogeography, Palaeoclimatology, Palaeoecology*, 192: 229–245.

Dahanayake, K., et al. 1985. Stromatolites, oncolites and oolites forming in situ. *Naturwissenschaften*, 72: 513–518.

Dattilo, B., et al. 2014. Stratigraphy of the Paluxy River tracksites in and around Dinosaur Valley State Park, Lower Cretaceous Glen Rose Formation, Somervell County, Texas. *New Mexico Museum of Natural History Bulletin*, 62: 307–338.

Davis, D.E. 2011. *Where There Are Mountains: An Environmental History of the Southern Appalachians*. University of Georgia Press, Athens: 353 p.

Debczak, M. 2016. 10 eye-popping facts about mantis shrimp. *Mental Floss*, September 22. http://mentalfloss.com/article/86128/10-eye-popping-facts -about-mantis-shrimp.

De Stoppelaire, G.H., et al. 2004. Use of remote sensing techniques to determine the effects of grazing on vegetation cover and dune elevation at Assateague Island National Seashore: Impact of horses. *Environmental Management*, 34: 642–649.

Dial, K.P., et al. 2006. What use is half a wing in the ecology and evolution of birds? *BioScience*, 56: 437–445.

Dietl, G.P. 2004. Origins and circumstances of adaptive divergence in whelk feeding behavior. *Palaeogeography, Palaeoclimatology, Palaeoecology*, 208: 279–291.

Dilsaver, L.M. 2004. *Cumberland Island National Seashore: A History of Conservation Conflict*. University of Virginia Press, Charlottesville: 323 p.

Dixon, C. 2017. Reviving a lost cane syrup. *Garden and Gun*, December 14. https:// gardenandgun.com/articles/reviving-lost-cane-syrup/.

Duncan, G.A. 1986. Burrows of *Ocypode quadrata* (Fabricus) as related to slopes of substrate surfaces. *Journal of Paleontology*, 60: 384–389.

Ekdale, A.A., and Bromley, R.G. 2001. A day and a night in the life of a cleft-foot clam: *Protovirgularia–Lockeia–Lophoctenium*. *Lethaia*, 34: 119–124.

Elbroch, M. 2003. *Mammal Tracks and Sign: A Guide to North American Species*. Stackpole Books, Mechanicsburg, Pa.: 779 p.

Elbroch, M., and Marks, E. 2001. *Bird Tracks and Sign: A Guide to North American Species*. Stackpole Books, Mechanicsburg, Pa.: 456 p.

Ellers, O. 1995a. Behavioral control of swash-riding in the clam *Donax variabilis*. *Biological Bulletin*, 189: 120–127.

Ellers, O. 1995b. Discrimination among wave-generated sounds by a swash-riding clam. *Biological Bulletin*, 189: 128–137.

Engeman, R.M., et al. 2016. Defending against disparate marine turtle nest predators: Nesting success benefits from eradicating invasive feral swine and caging nests from raccoons. *Oryx*, 50: 289–295.

Evans, H.E, and O'Neill, K.M. 2009. *The Sand Wasps: Natural History and Behavior*. Harvard University Press, Cambridge, Mass.: 360 p.

Falcon-Lang, H.J. 2000. Fire ecology of the Carboniferous tropical zone. *Palaeogeography, Palaeoclimatology, Palaeoecology*, 164: 339–355.

Farlow, J.O., et al. 2017. Trackways of the American crocodile (*Crocodylus acutus*), northwest Costa Rica: Implications for archosaurian ichnology. *Ichnos*, 24: 1–36.

Farrely, C.A., and Greenaway, P. 1994. Gas exchange through the lungs and gills in air-breathing crabs. *Journal of Experimental Biology*, 187: 113–130.

Fierstien, J.F., IV, and Rollins, H.B. 1987. Observations on intertidal organism associations of St. Catherines Island, Georgia: II. Morphology and distribution of *Littorina irrorata* (Say). *American Museum Novitates*, 2873: 1–31.

Fincher, G.T., and Woodruff, R.E. 1979. Dung beetles of Cumberland Island, Georgia (Coleoptera: Scarabaeidae). *Coleopterists Bulletin*, 33: 69–70.

Fraser, W.J., Jr. 2006. *Lowcountry Hurricanes: Three Centuries of Storms at Sea and Ashore.* University of Georgia Press, Athens: 319 p.

Fredericks, A.D. 2012. *Horseshoe Crabs: Biography of a Survivor.* Ruka, Washington, D.C.: 256 p.

Frey, R.W., and Basan, P.B. 1981. Taphonomy of relict Holocene salt marsh deposits, Cabretta Island, Georgia. *Senckenbergiana Maritima*, 13: 111–155.

Frey, R.W., and Howard, J.D. 1969. A profile of biogenic sedimentary structures in a Holocene barrier island-salt marsh complex, Georgia. *Transactions of the Gulf Coast Association Geological Society*, 19: 427–444.

Frey, R.W., and Pemberton, S.G. 1986. Vertebrate lebensspuren in intertidal and supratidal environments, Holocene barrier island, Georgia. *Senckenbergiana Maritima*, 18: 97–121.

Frey, R.W., et al. 1984. Tracemaking activities of crabs and their environmental significance: The ichnogenus *Psilonichnus*. *Journal of Paleontology*, 58: 333–350.

Fuller, E. 2002. *Dodo: From Extinction to Icon.* HarperCollins, New York: 180 p.

Galloway, J. 2015. Supermoon, rising sea levels put Tybee Island access under water. *Atlanta-Journal Constitution*, October 27. https://politics.myajc.com/blog/politics/supermoon-rising-sea-levels-put-tybee-island-access-under-water/kE3PD96bMnoXJDPSHtoJtO/.

Gandini, R., et al. 2010. Assinaturas icnológicas da sucessão sedimentar Rio Bonito no bloco central da jazida carbonífera de Iruí, Cachoeira do Sul (RS) [Ichnological signatures of the Rio Bonito sedimentary succession in the central block of the Iruí coal mine, Rio Grande do Sul (Brazil)]. *Gaea: Journal of Geoscience*, 1: 21–43 [in Portuguese with English translation of the abstract].

Gerrard, J.M., and Bortolotti, G.R. 2014. *The Bald Eagle: Haunts and Habits of a Wilderness Monarch.* Smithsonian Institute, Washington, D.C.: 192 p.

Gilman, E.F. (1999) 2007. *Borrichia frutescens, Sea Oxeye.* Document FPS69. Environmental Horticulture, Florida Cooperative Extension Service, Institute of Food and Agricultural Sciences, University of Florida, Gainesville: 2 p.

Gingras, M.K., et al. 2002. Resin cast of modern burrows provides analogs for composite trace fossils. *Palaios*, 17: 206–211.

Giuffra, E., et al. 2000. The origin of the domestic pig: Independent domestication and subsequent introgression. *Genetics*, 154: 1785–1791.

Gladwell, M. 2005. *Blink: The Power of Thinking without Thinking.* Back Bay Books, New York: 320 p.

Glasspool, I.J., and Scott, A.C. 2010. Phanerozoic concentrations of atmospheric oxygen reconstructed from sedimentary charcoal. *Nature Geoscience*, 3: 627–630.

Godfray, H.C.J. 1994. *Parasitoids: Behavioral and Evolutionary Biology*. Princeton University Press, Princeton: 473 p.

Goodloe, R.B., et al. 2000. Population characteristics of feral horses on Cumberland Island and their management implications. *Journal of Wildlife Management*, 64: 114–121.

Gormally, C.L., and Donovan, L.A. 2010. Responses of *Uniola paniculata* L. (Poaceae), an essential dune-building grass, to complex changing environmental gradients on the coastal dunes. *Estuaries and Coasts*, 33: 1237–1248.

Gould, E., et al. 1993. Function of the star in the star-shaped mole. *Journal of Mammalogy*, 37: 223–231.

Grant, J. 1981. A bioenergetic model of shorebird predation on infaunal amphipods. *Oikos*, 37: 53–62.

Greenaway, P., et al. 1984. The venous system of the terrestrial crab *Ocypode cordimanus* (Desmarest 1825) with particular reference to the vasculature of the lungs. *Journal of Morphology*, 181: 133–142.

Guthrie, A., et al. 2019. Sedimentary and ecological interactions of shoreline management and feral horses on Cumberland Island, Georgia. *Geological Society of America Abstracts with Programs*, 51(3). https://doi.org/10.1130/abs/2019SE-327143.

Haas, F. 1950. Hermit crabs in fossil snail shells in Bermuda. *Ecology*, 31: 152.

Hackney, A.D., et al. 2013. Mapping risk for nest predation on a barrier island. *Journal of Coastal Conservation*, 17: 615–621.

Halfpenny, J.C., and Bruchac, J. 2002. *Scats and Tracks of the Southeast: A Field Guide to the Signs of Seventy Wildlife Species*. Globe Pequot, Guilford, Conn.: 192 p.

Harding, J.M., et al. 2007. Comparison of predation signatures left by Atlantic oyster drills (*Urosalpinx cinerea* Say, Muricidae) and veined rapa whelks (*Rapana venosa* Valenciennes, Muricidae) in bivalve prey. *Journal of Experimental Marine Biology and Ecology*, 352: 1–11.

Harrington, T.C., et al. 2008. *Raffaelea lauricola*, a new ambrosia beetle symbiont and pathogen on the Lauraceae. *Mycotaxon*, 104: 399–404.

Harrington, T.C., et al. 2011. Isolations from the redbay ambrosia beetle, *Xyleborus glabratus*, confirm that the laurel wilt pathogen, *Raffaelea lauricola*, originated in Asia. *Mycologia*, 103: 1028–1036.

Harris, L.M., and Berry, D.R. 2014. *Slavery and Freedom in Savannah*, University of Georgia Press, Athens: 262 p.

Hartman, G. 1995. Age determination, age structure, and longevity in the mole *Scalopus aquaticus* (Mammalia: Insectivora). *Journal of Zoology*, 237: 107–122.

Heers, A.M., and Dial, K.P. 2012. From extant to extinct: Locomotor ontogeny and the evolution of avian flight. *Trends in Ecology and Evolution*, 27: 296–305.

Held, S., et al. 2005. Foraging behaviour in domestic pigs (*Sus scrofa*): Remembering and prioritizing food sites of different value. *Animal Cognition*, 8: 114–121.

Henry, V.J., Jr., et al. 1993. A regional overview of the geology of barrier complexes near Cumberland Island, Georgia. In: Farrell, K.M., et al. (editors), *Geomorphology and Facies Relationships of Quaternary Barrier Island Complexes Near St. Mary's Georgia. Georgia Geological Society Guidebook*, 13: 2–10.

Hess, H., et al. 2002. *Fossil Crinoids*. Cambridge University Press, New York: 275 p.

Hessenbruch, A. 2000. *Reader's Guide to the History of Science*. Taylor and Francis, Milton Park, U.K.: 934 p.

Hill, G.W., and Hunter, R.E. 1973. Burrows of the ghost crab *Ocypode quadrata* (Fabricus) on the barrier islands, south-central Texas Coast. *Journal of Sedimentary Research*, 43: 24–30.

Hobbs, H.H., Jr. 1981. *The Crayfishes of Georgia*. Smithsonian Institute Press, Washington, D.C.: 549 p.

Hoffecker, J.F. 2017. *Modern Humans: Their African Origin and Global Dispersal*. Columbia University Press, New York: 544 p.

Hooton N., et al. 2014. Survival and growth of planted *Uniola paniculata* and dune building using surrogate wrack on Perdido Key Florida, U.S.A. *Restoration Ecology*, 22: 710–707.

Horner, J.R. 1982. Evidence of colonial nesting and "site fidelity" among ornithischian dinosaurs. *Nature*, 297: 675–676.

Horner, J.R. 1984. The nesting behavior of dinosaurs. *Scientific American*, 250: 130–137.

Horner, J.R., and Gorman, J. 1988. *Digging Dinosaurs*. Workman, New York: 210 p.

Horner, J.R., and Makela, R. 1979. Nest of juveniles provides evidence of family structure among dinosaurs. *Nature*, 282: 296–298.

Hornocker, M., and Negri, S. (editors). 2009. *Cougar: Ecology and Conservation*. University of Chicago Press, Chicago: 304 p.

Howard, J.D., and Elders, C.A. 1970. Burrowing patterns of haustoriid amphipods from Sapelo Island, Georgia. In: Crimes, T.P., and Harper, J.C. (editors), *Trace Fossils*. Seel House Press, Liverpool: 243–262.

Hoyt, J.H., and Hails, J.R. 1967. Pleistocene shoreline sediments in coastal Georgia: Deposition and modification. *Science*, 155: 1541–1543.

Hoyt, J.H., and Henry, V.J., Jr. 1967. Influence of island migration on barrier-island sedimentation. *Geological Society of America Bulletin*, 78: 77–86.

Hoyt, J.H., et al. 1964. Late Pleistocene and recent sedimentation on the central Georgia coast, U.S.A. In: van Straaten, L.M.J.U. (editor), *Deltaic and Shallow Marine Deposits, Developments in Sedimentology I*. Elsevier, Amsterdam: 170–176.

Hulbert, R.C., Jr., and Pratt, A.E. 1998. New Pleistocene (Racholabrean) vertebrate faunas from coastal Georgia. *Journal of Vertebrate Paleontology*, 18: 412–429.

Hume, J.P. 2012. The dodo: From extinction to the fossil record. *Geology Today*, 28: 147–151.

Humprey, S.R. 1974. Zoogeography of the nine-banded armadillo (*Dasypus novemcinctus*) in the United States. *BioScience*, 24: 457–462.

Isenman, L. 2013. Understanding unconscious intelligence and intuition: *Blink* and beyond. *Perspectives in Biology and Medicine*, 56: 148–166.

Issa A., et al. 2018. Deaths related to Hurricane Irma: Florida, Georgia, and North Carolina, September 4–October 10, 2017. MMWR *Morbidity Mortality Weekly Report (CDC)*, 67(30): 829–832. https://doi.org/10.15585/mmwr.mm6730a5.

Jisan, M.A., et al. 2018. Hurricane Matthew (2016) and its impact under global warming scenarios. *Modeling Earth Systems and Environment*, 4: 97–109.

Johnson, M.N. 2009. *Sapelo Island's Hog Hammock*. Arcadia, Mount Pleasant, S.C.: 128 p.

Johnson, W.B. 2002. In the shadow of the revolution: Savannah's first generation of free African American elite in the new republic, 1790–1830. In: Finley, R., and Deblack, T. (editors), *The Southern Elite and Social Change: Essays in Honor of Willard B. Gatewood*. University of Arkansas Press, Fayetteville: 3–15.

Kabat, A.R. 1990. Predatory ecology of naticids gastropods with a review of shell boring predation. *Malacologia*, 32: 155–193.

Katz, L.C. 1980. Effects of burrowing by the fiddler crab *Uca pugnax* (Smith). *Estuarine Coastal Marine Science*, 11: 233–237.

Kawai, T., et al. 2015. *Freshwater Crayfish: A Global Overview*. CRC, Boca Raton, Fla.: 679 p.

Kelley, P.H., and Hansen, T.A. 2001. The role of ecological interactions in the evolution of naticid gastropods and their molluscan prey. In: Allmon, W.D., and Bottjers, D.J. (editors), *Evolutionary Paleoecology*, Columbia University Press, New York: 149–170.

Kendra, P.E., et al. 2013. An uncertain future for American Lauraceae: A lethal threat from redbay ambrosia beetle and laurel wilt disease (a review). *American Journal of Plant Sciences*, 4: 727–738.

Kitching, R.L., and Pearson, J. 1981. Prey localization by sound in a predatory intertidal gastropod. *Marine Biology Letters*, 2: 313–321.

Kouwenberg, A.-L., et al. 2009. Episodic-like memory in crossbred Yucatan minipigs (*Sus scrofa*). *Applied Animal Behaviour*, 117: 165–172.

Labandeira, C.C. 2013. Deep-time patterns of tissue consumption by terrestrial arthropod herbivores. *Naturwissenschaften*, 100: 355–364.

Laerm, J., et al. 1999. Amphibians, reptiles, and mammals of Sapelo Island, Georgia. *Journal of the Elisha Mitchell Scientific Society*, 115: 104–126.

Lamsdell, J.C. 2015. Horseshoe crab phylogeny and independent colonizations of fresh water: Ecological invasion as a driver for morphological innovation. *Palaeontology*, 59: 181–194.

Landers, M. 2014. Tidal flooding forecast as new normal for Savannah, Tybee Island. *Savannah Morning News (Savannah Now)*, October 18. https://www.savannahnow.com/news/2014-10-18/tidal-flooding-forecast-new-normal-savannah-tybee-island.

Landers, M. 2015. Snap the tide, see Tybee's future. *Savannah Morning News (Savannah Now)*, October 29. https://www.savannahnow.com/news/2013-10-29/snap-tide-see-tybees-future.

Landry, C.E., et al. 2003. An economic evaluation of beach erosion management alternatives. *Marine Resource Economics*, 18: 105–127.

Larson, G., et al. 2008. Current views on *Sus* phylogeny and pig domestication as seen through modern mtDNA studies. In: Albarella, U., et al. (editors), *Pigs and Humans: 10,000 Years of Interaction*. Oxford University Press, Oxford: 30–41.

Larson, L.H. 1991. *A Guide to the Archaeology of Sapelo Island, Georgia*. Carrollton: West Georgia College [University of West Georgia]. https://doi.org/10.6067/XCV8JM28NZ.

Legendre, F., et al. 2015. Phylogeny of Dictyoptera: Dating the origin of cockroaches, praying mantises and termites with molecular data and controlled fossil evidence. *PLOS ONE* 10(7): e0130127. https://doi.org/10.1371/journal.pone.0130127.

Letnic, M., et al. 2012. Could direct killing by larger dingoes have caused the extinction of the thylacine from mainland Australia? *PLOS ONE* 7(5): e34877. https://doi.org/10.1371/journal.pone.0034877.

Lieutier, F., et al. 2007. *Bark and Wood Boring Insects in Living Trees in Europe, a Synthesis*. Springer, Berlin: 570 p.

Linsley, D.L., et al. 2008. Stratigraphy and geologic evolution of St. Catherines Island, Georgia. In: Thomas, D.H. (editor), *Native American Landscapes of St. Catherines Island. Anthropological Papers of the American Museum of Natural History*, 88: 26–41.

Lloyd, H.S. (1924) 2013. *The Popular Cocker Spaniel: Its History, Strains, Pedigrees, Breeding, Kennel Management, Ailments, Exhibition, Show Points, and Elementary Training for Sport and Field Trials, with a List of Winning Dogs*. Read Books, London, 248 p.

Locker, M. 2017. How one Georgia island is fighting to keep a small red pea alive. *Southern Living*, August 25. https://www.southernliving.com/cultur/southern-foodways-alliance-geechee-red-pea.

Lockley, M.G., et al. 2015. First report of bird tracks (*Aquatilavipes*) from the
Cedar Mountain Formation (Lower Cretaceous), eastern Utah. *Palaeogeography,
Palaeoclimatology, Palaeoecology*, 420: 150–162.

Longshaw, M., and Stebbing, P. 2016. *Biology and Ecology of Crayfish*. CRC, Boca
Raton, Fla.: 375 p.

Lott, D.F. 2002. *American Bison: A Natural History*. University of California Press,
Berkeley: 229 p.

Luther, H.E., and Benzing, D.H. 2009. *Native Bromeliads of Florida*. Rowman and
Littlefield, Lanham, Md.: 144 p.

MacDonald, C. 2016. Hurricane Matthew's destructive path. *Risk Management*, 63:
10–11.

Mann, C.C. 2011. *1493: Uncovering the New World Columbus Created*. Knopf, New
York: 560 p.

Manucy, A.C. 1952. Tapia or tabby. *Journal of the Society of Architectural Historians*,
11: 32–33.

Maples, C.G., and West, R.R. 1989. *Lockeia*, not *Pelecypodichnus*. *Journal of
Paleontology*, 63: 694–696.

Marquardt, W.H. 2010. Shell mounds in the Southeast: Middens, monuments,
temple mounds, rings, or works? *American Antiquity*, 75: 551–570.

Martin, A.J. 2006a. A composite trace fossil of decapod and hymenopteran
origin from the Rice Bay Formation (Holocene), San Salvador, Bahamas. In:
Gamble, D., and Davis, R.L. (editors), *12th Symposium of the Geology of the
Bahamas and Other Carbonate Regions*, Gerace Research Center, San Salvador,
Bahamas: 99–112.

Martin, A.J. 2006b. Resting traces of *Ocypode quadrata* associated with hydration
and respiration: Sapelo Island, Georgia, USA. *Ichnos*, 13: 57–67.

Martin, A.J. 2013. *Life Traces of the Georgia Coast: Revealing the Unseen Lives of Plants
and Animals*. University of Georgia Press, Athens: 692 p.

Martin, A.J. 2017. *The Evolution Underground: Burrows, Bunkers, and the
Marvelous Subterranean World beneath Our Feet*. Pegasus Books, New York:
460 p.

Martin, A.J., and Rindsberg, A.K. 2007. Arthropod tracemakers of *Nereites*?
Neoichnological observations of juvenile limulids and their paleoichnological
applications. In: Miller, W.M., III (editor), *Trace Fossils: Concepts, Problems,
Prospects*. Elsevier, Amsterdam: 478–491.

Martin, A.J., and Rindsberg, A.K. 2011. Ichnological diagnosis of ancient storm-
washover fans, Yellow Banks Bluff, St. Catherines Island. In: Bishop, G.A., et al.
(editors), *Geoarchaeology of St. Catherines Island, Georgia*. Anthropological Papers
of the American Museum of Natural History, no. 94: 114–127.

Martin, A.J., and Varricchio, D.J. 2011. Paleoecological utility of insect trace fossils in dinosaur nesting sites of the Two Medicine Formation (Campanian), Choteau, Montana. *Historical Biology*, 23: 15–25.

Martin, A.J., et al. 2008. Fossil evidence in Australia for oldest known freshwater crayfish of Gondwana. *Gondwana Research*, 14: 287–296.

Martin, A.J., et al. 2012. A polar dinosaur-track assemblage from the Eumeralla Formation (Albian), Victoria, Australia. *Alcheringa: An Australasian Journal of Palaeontology*, 36: 171–188.

Martin, A.J., et al. 2014. Oldest known avian footprints from Australia: Eumeralla Formation (Albian), Dinosaur Cove, Victoria. *Palaeontology*, 57: 7–19.

Martin, A.J., et al. 2015. The ups and downs of *Diplocraterion* in the Glen Rose Formation (Albian), Texas (USA). *Geodinamica Acta*, 28: 101–119.

Mayfield, A.E., III, et al. 2008. Ability of the redbay ambrosia beetle (Coleoptera: Curculionidae: Scolytinae) to bore into young avocado (Lauraceae) plants and transmit the laurel wilt pathogen (*Raffaelea* sp.). *Florida Entomologist*, 91: 485–487.

Mayor, J.J., Jr., and Brisbin, I.L. 2008. *Wild Pigs in the United States: Their History, Comparative Morphology, and Current Status*. University of Georgia Press, Athens: 336 p.

McClain, W.R., and Romaire, R.P. 2004. Crawfish culture: A Louisiana aquaculture success story. *World Aquaculture*, 35: 31–35.

McInerney, F.A., and Wing, S.L. 2011. The Paleocene-Eocene Thermal Maximum: A perturbation of carbon cycle, climate, and biosphere with implications for the future. *Annual Review of Earth and Planetary Sciences*, 39: 489–516.

McTavish, E.J., et al. 2013. New World cattle show ancestry from multiple independent domestication events. *Proceedings of the National Academy of Sciences*, 110: E1398–E1406.

Meeker, J.R., et al. 1995. The Southern pine beetle *Dendroctonus frontalis* Zimmerman (Coleoptera: Scolytidae). Florida Department of Agricultural and Consumer Services. *Entomology Circular*, 369: 1–4.

Mikuláš, R. 2001. Modern and fossil traces in terrestrial lithic substrates. *Ichnos*, 8: 177–184.

Milanich, J.T. 1996. *The Timucua*. Peoples of America series. Wiley-Blackwell, Oxford: 256 p.

Monroe, K. 2014. The weeping time: A forgotten history of the largest slave auction ever on American soil. *Atlantic*, July 10. https://www.theatlantic.com/business/archive/2014/07/the-weeping-time/374159/.

Montalvo, C.I. 2002. Root traces in fossil bones from the Huayquerian (Late Miocene) faunal assemblage of Telén, La Pampa, Argentina. *Acta Geológica Hispánica*, 37: 37–42.

Morris, R.W., and Rollins, H.B. 1977. Observations on intertidal organism associations on St. Catherines Island, Georgia: I. General description and paleoecological implications. *Bulletin of the American Museum of Natural History*, 159: 87–128.

Müller, A.H. 1956. Über problematische Lebensspuren aus dem Rotliegenden von Thüringen. *Berichte der Geologischen Gesellschaft in der Deutschen Demokratischen Republik für das Gesamtgebiet der Geologischen Wissenschaften*, 1: 147–155.

Negreiros-Fransozo, M.L., et al. 2002. Reproductive cycle and recruitment period of *Ocypode quadrata* (Decapoda, Ocypodidae) at a sandy beach in southeastern Brazil. *Journal of Crustacean Biology*, 22: 157–161.

Nel, A., et al. 2018. Palaeozoic giant dragonflies were hawker predators. *Scientific Reports*, 8: 12141: https://www.nature.com/articles/s41598-018-30629-w.

Nichols, G. 2012. *Sedimentology and Stratigraphy* (2nd Edition). Wiley-Blackwell, Oxford: 456 p.

Odum, E.P. 1968. Energy flow in ecosystems: A historical review. *American Zoologist*, 8: 11–18.

Ogburn, S.P. 2013. Ice-free Arctic in Pliocene, last time CO_2 levels above 400 PPM. *Scientific American*, May 10. https://www.scientificamerican.com/article/ice-free-arctic-in-pliocene-last-time-co2-levels-above-400ppm/.

Okajima, R. 2008. The controlling factors limiting maximum body size of insects. *Lethaia*, 41: 423–430.

Pasini, G., et al. 2016. Anomuran and brachyuran trackways and resting trace from the Pliocene of Valduggia (Piedmont, NW Italy): Environmental, behavioural, and taphonomic implications. *Natural History Sciences*, 3: 35–48.

Pejchar, L., and Mooney, H.A. 2009. Invasive species, ecosystem services and human well-being. *Trends in Ecology and Evolution*, 24: 497–504.

Peris, D., et al. 2014. The earliest occurrence and remarkable stasis of the family Bostrichidae (Coleoptera: Polyphaga) in Cretaceous Charentes amber. *Palaeontologia Electronica*, 17: 1–8.

Pervesler, P., and Uchman, A. 2009. A new Y-shaped trace fossil attributed to upogebiid crustaceans from Early Pleistocene of Italy. *Acta Palaeontologica Polonica*, 54: 135–142.

Petersen, K.E., and Yates, T.L. 1980. *Condylura cristata. Mammalian Species*, 129: 1–4.

Peterson, C.H. 1982. Clam predation by whelks (*Busycon* spp.): Experimental tests of the importance of prey size, prey density, and seagrass cover. *Marine Biology*, 66: 159–170.

Pickerell, J. 2014. *Flying Dinosaurs: How Fearsome Reptiles Became Birds*. NewSouth, Sydney: 256 p.

Pilkey, O., and Fraser, M.E. 2003. *A Celebration of the World's Barrier Islands*. Columbia University Press, New York: 309 p.

Pilkey, O.H., et al. 2017. *The World's Beaches: A Global Guide to the Science of the Shoreline*. University of California Press, Berkeley: 302 p.

Pirkle, W.A., and Pirkle, F.L. 2007. Introduction to heavy-mineral sand deposits of the Florida and Georgia Atlantic coastal plain. In: Rich, F.J. (editor), *Guide to Fieldtrips: 56th Annual Meeting, Southeastern Section of the Geological Society of America*. Georgia Southern University, Department of Geology and Geography Contribution, Series 1: 129–135.

Potter, P.E., et al. 1980. *Sedimentology of Shale: Study Guide and Reference Source*. Springer, Berlin: 310 p.

Prezant, R.S., et al. 2002. Marine macroinvertebrate diversity of St. Catherines Island, Georgia. *American Museum Novitates*, 3367: 1–31.

Proffitt, C.E., et al. 2003. Genotype and elevation influence *Spartina alterniflora* colonization and growth in a created salt marsh. *Ecological Applications*, 13: 180–192.

Quammen, D. 1996. *Song of the Dodo: Island Biogeography in an Age of Extinctions*. Scribner, New York: 702 p.

Quicke, D.L.J. 2014. *The Braconid and Ichneumonid Parasitoid Wasps: Biology, Systematics, Evolution and Ecology*. Wiley and Sons, New York: 688 p.

Raab, J.W. 2007. *Spain, Britain and the American Revolution in Florida, 1763–1783*. McFarland, Jefferson, N.C.: 210 p.

Rabaglia, R.J., et al. 2006. Review of American Xyleborina (Coleoptera: Curculionidae: Scolytinae) occurring north of Mexico, with an illustrated key. *Annals of the Entomological Society of America*, 99: 1034–1056.

Rehan, S.M. 2012. A Mid-Cretaceous origin of sociality in xylocopine bees with only two origins of true worker castes indicates severe barriers to eusociality. *PLOS ONE*, 7: e34690. https://doi.org/10.1371/journal.pone.0034690.

Reineck, H.-E. 1981. Lebensspur of a bird starting to fly in snow on the German–Austrian border. *Journal of Sedimentary Petrology*, 51: 699.

Reolid, J., and Reolid, M. 2017. Traces of floating archosaurs: An interpretation of the enigmatic trace fossils from the Triassic of the tabular cover of southern Spain. *Ichnos*, 24: 222–233.

Rich, T.H., and Vickers-Rich, P. 2000. *Dinosaurs of Darkness*. Indiana University Press, Bloomington: 222 p.

Rijsdijk, K.F., et al. 2015. A review of the dodo and its ecosystem: Insights from a vertebrate concentration lagerstätte in Mauritius. *Journal of Vertebrate Paleontology*, 35 (supplement 1): 3–20. https://doi.org/10.1080/02724634.2015 .1113803.

Rindsberg, A.K., and Martin, A.J. 2003. *Arthrophycus* and the problem of compound trace fossils. *Palaeogeography, Palaeoclimatology, Palaeoecology*, 192: 187–219.

Rindsberg, A.K., and Martin, A.J. 2015. Caster's plasters: Neoichnological experiments by Kenneth Caster on limulids in 1937. *Geological Association of Canada*. Special Volume 9, Papers from *Ichnia*. International Ichnological Congress Meeting, St. Johns, Newfoundland (Canada): 197–210.

Robertson, J.R., and Pfeiffer, W.J. 1982. Deposit feeding by the ghost crab *Ocypode quadrata*. *Journal Experimental Marine Biology and Ecology*, 56: 165–177.

Rogers, R.R., et al. 2010. *Bone Beds: Genesis, Analysis, and Paleobiological Significance*. University of Chicago Press, Chicago: 512 p.

Rokosz, M. 1995. History of the aurochs (*Bos taurus primigenius*) in Poland. *Animal Genetics Resources Information, Food and Agriculture Organization*, 16: 5–12.

Rosenblatt, A.E., and Heithaus, M.R. 2015. Does variation in movement tactics and trophic interactions among American alligators create habitat linkages? *Journal of Animal Ecology*, 80: 786–798.

Ruckdeschel, C. 2017. *A Natural History of Cumberland Island, Georgia*. Mercer University Press, Macon: 376 p.

Ruckdeschel, C., and Shoop, C.R. 2012. *Sea Turtles of the Atlantic and Gulf Coasts of the United States*. University of Georgia Press, Athens: 152 p.

Rudkin, D.M., et al. 2008. The oldest horseshoe crab: A new xiphosurid from Late Ordovician Konservat-Lagerstätten deposits, Manitoba, Canada. *Palaeontology*, 51: 1–9.

Ruppert, E.E., and Fox, R.S. 1988. *Seashore Animals of the Southeast*. University of South Carolina Press, Columbia: 429 p.

Russo, M. 2006. *Archaic Shell Rings of the Southeast U.S.: National Historic Landmarks Historic Context*. Southeast Archeological Center, U.S. National Park Service, Tallahassee: 173 p.

Sabine, J.B., et al. 2006. Nest fate and productivity of American Oystercatchers, Cumberland Island National Seashore, Georgia. *Waterbirds*, 29: 309–316.

Sakai, K. 2005. Callianassoidea of the World (Decapoda, Thalassinidea). *Crustaceana Monographs*, 4: 1–285.

Sanger, M.C., and Thomas, D.H. 2010. The two rings of St. Catherines Island: Some preliminary results from the St. Catherines and McQueen Shell rings. In: Hurst, D.H., and Sanger, M.C. (editors), *Trend, Tradition, and Turmoil: What Happened to the Southeastern Archaic?* American Museum of Natural History Anthropological Papers, no. 93: 45–70.

Sanger, M.E. 2015. Determining depositional events within shell deposits using computer vision and photogrammetry. *Journal of Archaeological Science*, 53: 482–491.

Sassaman, K.E. 1993. *Early Pottery in the Southeast: Tradition and Innovation in Cooking Technology.* University of Alabama Press, Tuscaloosa: 312 p.

Schaefer, J. (translator). 1989. *Recent Vertebrate Carcasses and Their Paleobiological Implications.* By A. Weigelt. University of Chicago Press, Chicago: 204 p.

Schnakenberg, H. 2010. *Kid Carolina: R. J. Reynolds Jr., a Tobacco Fortune, and the Mysterious Death of a Southern Icon,* Center Street, New York: 352 p.

Schweigert, G. 2011. The decapod crustaceans of the Upper Jurassic Solnhofen Limestones: A historical review and some recent discoveries. *Neues Jahrbuch für Geologie und Paläontologie: Abhandlungen,* 260: 131–140.

Scott, J.T. 1994. The Frederica Homefront in 1742. *Georgia Historical Quarterly,* 78: 493–508.

Seibold, E., and Berger, W. 2017. *The Sea Floor: An Introduction to Marine Geology.* Springer, Berlin: 268 p.

Sharp, S.J., and Angelini, C. 2016. Whether disturbances alter salt marsh soil structure dramatically affects *Spartina alterniflora* recolonization rate. *Ecosphere,* 7: e01540.

Sherr, E.B. 2015. *Marsh Mud and Mummichogs: An Intimate Natural History of Coastal Georgia.* University of Georgia Press, Athens: 256 p.

Shuster, C.N., Jr., et al. 2003. *The American Horseshoe Crab.* Harvard University Press, Cambridge, Mass.: 427 p.

Sickels-Taves, L.B., and Sheehan, M.S. 1999. *The Lost Art of Tabby Redefined: Preserving Oglethorpe's Architectural Legacy.* Architectural Conservation, Southfield, Mich.: 200 p.

Sickels-Taves, L.B., and Sheehan, M.S. 2002. Specifying historic materials: The use of lime. In: Throop, D., and Klingner, R.E. (editors), *Masonry: Opportunities for the 21st Century.* ASTM International, West Conshohocken, Pa.: 3–22.

Silliman, B.R., and Bertness, M.D. 2002. A trophic cascade regulates salt marsh primary production. *Proceedings of the National Academy of Sciences,* 99: 10500–10505.

Skelton, C.E. 2010. History, status, and conservation of Georgia crayfishes. *Southeastern Naturalist,* 9: 127–138.

Smith, B., and Savolainen, P. 2015. The origin and ancestry of the dingo. In: Smith, B. (editor), *The Dingo Debate: Origins, Behaviour, and Conservation.* CSIRO, Clayton, Australia: 55–80.

Smith, C.K., and McGrath, D.A. 2011. The alteration of soil chemistry through shell deposition on a Georgia (U.S.A.) barrier island. *Journal of Coastal Research,* 27: 103–109.

Smith, D., et al. 2015. The early Holocene sea level rise. *Quaternary Science Reviews,* 30: 1846–1860.

Smith, J.M., and Frey, R.W. 1985. Biodeposition by the ribbed mussel *Geukensia demissa* in a salt marsh, Sapelo Island, Georgia. *Journal of Sedimentary Research*, 55: 817–825.

Smith, S.E., and Read, D.J. 2008. *Mycorrhizal Symbiosis*. Academic Press, Cambridge, Mass.: 880 p.

Sobkowiak, S., et al. 1989. Bald eagles killing American coots and stealing coot carcasses from greater black-backed gulls. *Wilson Bulletin*, 101: 494–496.

Spiegel, K.S., and Leege, L.M. 2013. Impacts of laurel wilt disease on redbay (*Persea borbonia* [L.] Spreng.) population structure and forest communities in the coastal plain of Georgia, USA. *Biological Invasions*, 15: 2467–2487.

Stewart, M.A. 2002. *"What Nature Suffers to Groe": Life, Labor, and Landscape on the Georgia Coast, 1680–1920*. University of Georgia Press, Athens: 392 p.

Stuart, A.J. 2017. Late Quaternary megafaunal extinctions on the continents: A short review. *Geological Journal*, 50: 338–363.

Subramoniam, T. 2016. *Sexual Biology and Reproduction in Crustaceans*. Academic Press, New York: 526 p.

Sues, H.-D., and Fraser, N. 2010. *Triassic Life on Land: The Great Transition*. Columbia University Press, New York: 280 p.

Tapanila, L., and Roberts, E.M. 2012. The earliest evidence of holometabolan insect pupation in conifer wood. *PLOS ONE* 7: e31668. https://doi.org/10.1371/journal.pone.0031668.

Taylor, J.W., et al. 2004. The Fungi. In: Cracraft, J., and Donoghue, M.J. (editors), *Assembling the Tree of Life*. Oxford University Press, Oxford: 171–193.

Taylor, R.B., et al. 1998. Reproduction of feral pigs in southern Texas. *Journal of Mammalogy*, 79: 1325–1331.

Teal, J.M. 1958. Distribution of fiddler crabs in Georgia salt marshes. *Ecology*, 39: 186–193.

Teal, J.M. 1962. Energy flow in the salt marsh ecosystem of Georgia. *Ecology*, 43: 614–624.

Teal, M., and Teal, J.M. 1964. *Portrait of an Island*. Atheneum, New York: 167 p. University of Georgia Press, Athens, 1997, 184 p.

Thaler, A.D. 2016. The politics of fake documentaries. *Slate*, August 13. https://slate.com/technology/2016/08/the-lasting-damage-of-fake-documentaries-like-mermaids-the-body-found.html.

Thompson, V.D. 2007. Articulating activity areas and formation processes at the Sapelo Island shell ring complex. *Southeastern Archaeology*, 26: 91–107.

Thompson, V.D., and J.A. Turck. 2010. Island archaeology and the Native American economies (2500 B.C.–A.D. 1700) of the Georgia coast. *Journal of Field Archaeology*, 35: 283–297.

Thompson, V.D., et al. 2004. The Sapelo Island shell ring complex: Shallow geophysics on a Georgia sea island. *Southeastern Archaeology*, 23: 192–201.

Trueman, E.R., and Brown, A.C. 1992. The burrowing habit of marine gastropods. *Advances in Marine Biology*, 28: 389–431.

Trueman, E.R., et al. 1966. The dynamics of burrowing of some common littoral bivalves. *Journal of Experimental Biology*, 44: 469–492.

Turck, J.A., and Thompson, V.D. 2016. Revisiting the resilience of Late Archaic hunter-gatherers along the Georgia coast. *Journal of Anthropological Archaeology*, 43: 39–55.

Turner, H.J., Jr., and Belding, D.L. 1957. The tidal migrations of *Donax variabilis* Say. *Limnology and Oceanography*, 2: 120–124.

Turner, M.G. 1987. Effects of grazing by feral horses, clipping, trampling, and burning on a Georgia salt marsh. *Estuaries and Coasts*, 10: 54–60.

Turner, M.G. 1988. Simulation and management implications of feral horse grazing on Cumberland Island, Georgia. *Journal of Range Management*, 41: 441–447.

Turra, A., et al. 2005. Predation on gastropods by shell-breaking crabs: Effects on shell availability to hermit crabs. *Marine Ecology Progress Series*, 286: 279–291.

U.S. Army Corps of Engineers. 2018. "Tybee Island Gets Renourished." *Balancing the Basin*, April 16. http://balancingthebasin.armylive.dodlive.mil/2018/04/16/tybee-island-gets-renourished/.

Vanderplank, S.E., et al. 2014. Biodiversity and archeological conservation connected: Aragonite shell middens increase plant diversity. *BioScience*, 64: 202–209.

Van der Wal, C., et al. 2017. The evolutionary history of Stomatopoda (Crustacea: Malacostraca) inferred from molecular data. *PeerJ*, 5: e3844. https://peerj.com/articles/3844/.

Van Valen, L. 1973. A new evolutionary law. *Evolutionary Theory*, 1: 1–30.

Van Valen, L. 1974. Two modes of evolution. *Nature*, 252: 298–300.

Van Vuure, T. 2005. *Retracing the Aurochs: History, Morphology and Ecology of an Extinct Wild Ox*. Pensoft, Sophia, Bulgaria: 431 p.

Van Vuure, T. 2014. Aurochs *Bos primigenius* bojanus 1827. In: Melletti, M., and Burton, J. (editors), *Ecology, Evolution and Behaviour of Wild Cattle*. Cambridge University Press, New York: 240–254.

Varricchio, D.J., et al. 1999. A nesting trace with eggs for the Cretaceous theropod dinosaur *Troodon formosus*. *Journal of Vertebrate Paleontology*, 19: 91–100.

Veevers, J.J. 2006. Updated Gondwana (Permian-Cretaceous) earth history of Australia. *Gondwana Research*, 9: 231–260.

Vega, F.E., and Hofstetter, R.W. (editors). 2014. *Bark Beetles: Biology and Ecology of Native and Invasive Species*. Academic Press, Cambridge, Mass.: 640 p.

Vermeij, G.J. 1982. Gastropod shell form, breakage, and repair in relation to predation by the crab *Calappa*. *Malacologica*, 23: 1–12.

Visaggi, C.C., et al. 2013. Testing the influence of sediment depth on drilling behaviour of *Neverita duplicata* (Gastropoda: Naticidae), with a review of alternative modes of predation by naticids. *Journal of Molluscan Studies*, 79: 310–322.

Walker, S.E. 1989. Hermit crabs as taphonomic agents. *Palaios*, 4: 439–452.

Walker, S.E. 1992. Criteria for recognizing marine hermit crabs in the fossil record using gastropod shells. *Journal of Paleontology*, 66: 535–558.

Wang, J.Q, et al. 2010. Bioturbation of burrowing crabs promotes sediment turnover and carbon and nitrogen movements in an estuarine salt marsh. *Ecosystems*, 13: 586–599.

Wang, X., and Tedford, R.H. 2008. *Dogs: Their Fossil Relatives and Evolutionary History*. Columbia University Press, New York: 232 p.

Weigelt, A. 1927. *Rezente Wirbeltierleichen und ihre Paläobiologische*. Bedeutung. Max Weg, Leipzig, Germany: 227 p.

Weimer, R.J., and Hoyt, J.H. 1964. Burrows of *Callianassa major* Say, geologic indicators of littoral and shallow neritic environments. *Journal of Paleontology*, 38: 761–767.

Wilson, J. 2011. *Common Birds of Coastal Georgia*. University of Georgia Press, Athens: 219 p.

Witherington, B., and Witherington, D. 2011. *Living Beaches of Georgia and the Carolinas: A Beachcombers Guide*. Pineapple, Sarasota, Fla.: 342 p.

Wolcott, T.G. 1978. Ecological role of ghost crabs, *Ocypode quadrata* (Fabricius) on an ocean beach: Scavengers or predators? *Journal of Experimental Marine Biology and Ecology*, 31: 67–82.

Wolcott, T.G. 1984. Uptake of interstitial water from soil: Mechanisms and ecological significance in the ghost crab *Ocypode quadrata* and two gecarcinid land crabs. *Physiological Zoology*, 57: 161–184.

Wood, G.W., and Roark, D.N. 1980. Food habits of feral hogs in coastal South Carolina. *Journal of Wildlife Management*, 44: 506–511.

Woodruff, A. 2014. America's original razorbacks. *Twilight Beasts*, July 12. https://twilightbeasts.org/2014/07/12/americas-original-razorbacks/.

World Register of Marine Species. 1795. *Ocypode Weber*. http://www.marinespecies.org/aphia.php?p=taxdetails&id=106970.

Worth, J.E. 2007. *The Struggle for the Georgia Coast*. University of Alabama Press, Tuscaloosa: 222 p.

Yang, B., et al. 2012. Use of LiDAR shoreline extraction for analyzing revetment rock beach protection: A case study of Jekyll Island State Park, USA. *Ocean and Coastal Management*, 69: 1–15.

Yefremov, J. A. 1940. Taphonomy: New branch of paleontology. *Pan American Geologist*, 74: 81–93.

Zannuttigh, B., et al. 2014. *Coastal Risk Management in a Changing Climate*. Butterworth-Heinemann (Elsevier), Oxford, U.K.: 670 p.

Zipser, E., and Vermeij, G.J. 1978. Crushing behavior of tropical and temperate crabs. *Journal of Experimental Marine Biology and Ecology*, 31: 155–172.

Zomlefer, W.B., et al. 2008. Vascular plant survey of Cumberland Island National Seashore, Camden County, Georgia. *Castanea*, 73: 251–282.

INDEX

cats: feral, 90, 151; saber-toothed, 159
cattle, 151, 155–165, *160*, 252n5; tracks of,
    156–157, *161*, 172, 226
cement, 18, 19, 48–49, 221; of tabby ruins,
    199–202, *200*, *203*, 208–209
Cervus canadensis, 206–207
Charadrius melodus, 93
chelipeds, 52, 80
Chelonia mydas, 180–181
Chin, Karen, 194
Chocolate (Chucalate) Plantation, 199–201,
    200
cicadas, 89, 230, 231
clams, 12–13; coquina, 45–50, *47*, 167; dwarf
    surf, *4*, 6–8, *7*, 46; fiddler crabs and, 38;
    moon snails and, 71, 73–74; trace fossils of,
    113–114; whelks and, 7–8
claystone, 18
climate change, 154, 234–237, 258n6; high
    tides and, 215–217, *218–219*, 227, 233;
    intergovernmental panel on, 258n8;
    invasive species and, 197, 234; Pleistocene,
    84
cockroaches, 194–195
Columbina passerina, 118
Condylura cristata, 136–137
coprolites, 176, 187, 194
coquina clams, 45–50, *47*, 167
cordgrass, salt-meadow, 253n13. *See also*
    smooth cordgrass
cougars, 206
crabs, 22, 58; blue, 54, 75; European green, 151;
    hermit, 22, 38, *39*, 76–77; horseshoe, 63–70,
    66, *68–69*, 186, 233; moon snails and, 75;
    stone, 75. *See also* fiddler crabs; ghost crabs
Crassostrea virginica, 27, 28–30
Crawford, John "Crawfish," 126–127, 180
crayfish, 22, 58, 79–85, *81*, 167; red swamp,
    248n12; spiny, 83
crickets, 131
crinoids, 212–213, *214*
Cumberland Island, 17; ancient barrier islands
    and, *20*; cattle on, 252n5; coquina clams on,
    45–50, *47*; crayfish on, *81*, 84; ecosystems
    of, 192; ferry to, 166; hogs on, 180, 183, 187,
    188; horses on, 45, 158, 166–178, *168–169*,
    174–175; Spanish moss on, 173

Dasypus novemcinctus, 151–154, *152–153*
decapods, 22, 58, 82
deer, 172, 206–207; hogs and, 183, 186–187;
    white-tailed, 162, 187

Dendroctonus frontalis, 197, 226
Dermochelys coriacea, 181
diagenesis, 37
diamondback terrapin, 181
dingo, 150, 251nn3–4
dinosaurs: birds as, 117, 124, 128; at Egg
    Mountain, 91–92; fossils of tracks and
    bones, 121, 124; insects and, 92–93, 95, 194;
    nesting of, 91–93, 95; parenting habits,
    91–92; timescale of, 35, 233
Dinosaur Valley State Park (Tex.), 32–33
diversity. *See* biodiversity
dodo bird, 181
Donax variabilis, 45–50, *47*, 167
doves, 118
dragonflies, 21
dung beetles, 176, 254n15
dung flies, 176
Dutch elm disease, 193
dwarf surf clams. *See under* clams

eagles, 128–129, 166
Earle, Steven, 244n6
eastern oysters, 27, 28–30
echinoderms, 212–213
ecosystems, 15, 24, 35, 192; hogs and, 181;
    horses and, 173–178; invasive species and,
    149–154, *152–153*; moon snails and, 72;
    "pristine," 171, 179
egrets, 116–118, *119*, *123*
Ekdale, Tony, 113–114
elk, 206–207
erosion rates, 232
European green crabs, 151

Felis catus, 90, 151
fiddler crabs, 54, 99, *100*, 233; mud, 27, 28–30,
    38, *39*, 221, 236; sand, 101–103, *102*, 221, 231
firmgrounds, 229, 230
flies, dung, 176
flight tracks, 116–125, *119*, *123*
flood tides, 215–217, *218–219*
forests, pine, 23, 197. *See also* maritime forests
Fort Frederica, 209, 257n22
fossils: in ballast stones, *214*; body, 33–34; of
    brachiopods, 74; of crinoids, 212–213, *214*;
    of feces, 176, 187; of flight tracks, 116–118,
    120–125, *123*; of frass, 194; of fungi, 195; of
    hogs, 187; of horses, 171; taphonomy of,
    34–37. *See also* trace fossils
frass, 194, *195*
Frey, Robert, 72, 135, 136

maritime forests, 173, 234; mole burrow in, 132; redbay in, 191, 196; storm-washover fans and, 100

marshes. *See* salt marshes

Menippe mercenaria, 75

Mercenaria mercenaria, 202

metamorphic rocks, 16, 87, 212

Miami University (Ohio), 25, 244n1

mice, 110–112, 131–132

middens, 204–209

Milankovitch cycles, 258n7

Milton, John, 232, 259n6

Mocama people, 204

moles, 130–138, *132–134*

moon snails, 71–78, *76, 77,* 233; burrows of, *72–73,* 248n21; larvae of, 78

mosquito-borne diseases, 186

mud fiddler crabs. *See under* fiddler crabs

Mulinia lateralis, 4, 6–8, *7,* 46

muricids, 75

muscadine vines, 201

mussels, ribbed, 28–30, 38–39, *40,* 236

Mylohyus nasutus, 182

Nannygoat Beach, 51, 157, 215, 226; ghost crabs on, 54–55, *55,* 130. *See also* Sapelo Island

naticids. *See* gastropods

Neverita duplicata. See moon snails

New Jersey barrier islands, 26

nor'easter storms, 207

nutrient cycling, 15

oak. *See* live oak

Occam's razor, 135, 251n3

octopus, 75

Ocypode quadrata. See ghost crabs

Odingsell, Anthony, 179

Odocoileus virginianus, 162, 187

Odum, Eugene, 26

Oglethorpe, James Edward, 209, 257n22

oncolites, 213

operculum, 72

Ophiostoma fungi, 193

opisthosoma, 64–67, *66*

ornithomimid, 124

ornithopods, 121

Ossabaw Island, 17; ancient barrier islands and, *20*; hogs on, 180, 182, 183, 187; horses on, 180; after Hurricane Matthew, 220, 222–223; shell rings on, 204

otters, river, 139–145, *142–144*

oviraptorid, 124

owls, 118

oyster drill, Atlantic, 75

oystercatcher, American, 93

oysters, 236; eastern, 27, 28–30; in relict marshes, 39–40; in tabby ruins, *200*

Page, Michael, 100, 215–216

Page, Rebecca, 215

Palaeoechinastacus australanus, 83

palmettos, saw, 158, 192, 202

Pamlico (ancient shoreline), 19, *20*

parasitoid behavior, 89–90

Parsons family, 179–180

peccary, 182, 255n15

pelicans, 118, 120, 166

Pemberton, George, 136

Penholoway (ancient shoreline), 19, *20*

periwinkles, 28–30, 35, 37–39, 103, 236

Peromyscus polionotus, 110–112, 131–132

Persea americana, 196

Persea borbonia, 191–197, *195*

phytoplankton, 78

pine forests, 23, 197. *See also* maritime forests

Plato, 139

Platygonus compressus, 182

plovers, 93, 119

Poe, Edgar Allen, 72

polychaete worms, 213, 233

Pomeroy, Lawrence "Larry," 26

Princess Anne (ancient shoreline), 19, *20*

Procambarus clarkii, 248n12

Procyon lotor, 100, *102,* 103–106, 181

prosoma, limulid, 64, *66, 67*

Pterois volitans, 151

Puma concolor, 206

quahog, southern, 202

quartz, 16

quartz sand, 17, 18

Quercus virginiana. See live oak

raccoons, 100, *102,* 103–106, 181

radula, 5, 73

Raffaelea lauricola, 151, 193, *195,* 196–197

Rana sphenocephala, 112

Raphus cucullatus, 181

rats, 181

red knots, 65

Red Queen hypothesis, 128–129

red swamp crayfish, 248n12

redbay, 191–197, *195*